Advances in Intelligent Systems and Computing

Volume 1008

The series "Advances in Intelligent Systems and Computing" contains publications on theory, applications, and design methods of Intelligent Systems and Intelligent Computing. Virtually all disciplines such as engineering, natural sciences, computer and information science, ICT, economics, business, e-commerce, environment, healthcare, life science are covered. The list of topics spans all the areas of modern intelligent systems and computing such as: computational intelligence, soft computing including neural networks, fuzzy systems, evolutionary computing and the fusion of these paradigms, social intelligence, ambient intelligence, computational neuroscience, artificial life, virtual worlds and society, cognitive science and systems, Perception and Vision, DNA and immune based systems, self-organizing and adaptive systems, e-Learning and teaching, human-centered and human-centric computing, recommender systems, intelligent control, robotics and mechatronics including human-machine teaming, knowledge-based paradigms, learning paradigms, machine ethics, intelligent data analysis, knowledge management, intelligent agents, intelligent decision making and support, intelligent network security, trust management, interactive entertainment, Web intelligence and multimedia.

The publications within "Advances in Intelligent Systems and Computing" are primarily proceedings of important conferences, symposia and congresses. They cover significant recent developments in the field, both of a foundational and applicable character. An important characteristic feature of the series is the short publication time and world-wide distribution. This permits a rapid and broad dissemination of research results.

** Indexing: The books of this series are submitted to ISI Proceedings, EI-Compendex, DBLP, SCOPUS, Google Scholar and Springerlink **

More information about this series at http://www.springer.com/series/11156

Elvira Popescu · Ana Belén Gil ·
Loreto Lancia · Luigia Simona Sica ·
Anna Mavroudi
Editors

Methodologies and Intelligent Systems for Technology Enhanced Learning, 9th International Conference, Workshops

 Springer

Editors
Elvira Popescu
Faculty of Automation, Computers
and Electronics, Computers
and Information Technology Department
University of Craiova
Craiova, Romania

Loreto Lancia
Department of Life, Health
and Environmental Sciences
University of L'Aquila
L'Aquila, L'Aquila, Italy

Anna Mavroudi
KTH Royal Institute of Technology
Stockholm, Sweden

Ana Belén Gil
BISITE Digital Innovation Hub
University of Salamanca
Salamanca, Spain

Luigia Simona Sica
Dipartimento di Studi umanistici
Univeriosta degli Studi die Napoli
Federico II
Naples, Italy

ISSN 2194-5357   ISSN 2194-5365   (electronic)
Advances in Intelligent Systems and Computing
ISBN 978-3-030-23883-4   ISBN 978-3-030-23884-1   (eBook)
https://doi.org/10.1007/978-3-030-23884-1

This Springer imprint is published by the registered company Springer Nature Switzerland AG
The registered company address is: Gewerbestrasse 11, 6330 Cham, Switzerland

Preface

Education is the cornerstone of any society, and it serves as one of the foundations for many of its social values and characteristics. Different methodologies and intelligent technologies are employed for creating Technology Enhanced Learning (TEL) solutions. Solutions are intelligent when they are rooted in artificial intelligence, stand-alone, or interconnected to others. They target not only cognitive processes but also motivational, personality, or emotional factors. In particular, recommendation mechanisms enable us to tailor learning to different contexts and people, e.g., by considering their personality. The use of learning analytics also helps us augment learning opportunities for learners and educators alike, e.g., learning analytics can support self-regulated learning or adaptation of the learning material. Besides technologies, methods help create novel TEL opportunities. Methods come from different fields, such as educational psychology or medicine, and from diverse communities co-working with people, such as making communities and participatory design communities. Methods and technologies are also used to investigate and enhance learning for "fragile users," like children, elderly people, or people with special needs.

This year's technical program of MIS4TEL conference presents both high quality and diversity, with contributions in well-established and evolving areas of research. The program features also three selected workshops, which aim to provide participants with the opportunity to present and discuss novel research ideas on emerging topics complementing the main conference. In particular, the workshops focus on multi-disciplinary and transversal aspects like TEL in nursing education programs, TEL in digital creativity education, and student assessment and learning design evaluation in TEL systems. A total of 19 quality papers, with authors coming from various European countries, have been selected for the workshops and included in the present volume.

We would like to thank the sponsors (IEEE Systems Man and Cybernetics Society Spain Section Chapter and the IEEE Spain Section (Technical Co-Sponsor), IBM, Indra, Viewnext, Global Exchange, AEPIA and APPIA), and finally, the

Local Organization members and the Program Committee members for their hard work, which was essential for the success of MIS4TEL'19.

Elvira Popescu
Ana Belén Gil
Loreto Lancia
Luigia Simona Sica
Anna Mavroudi

Organization of MIS4TEL 2019

http://www.mis4tel-conference.net/

General Chair

Rosella Gennari   Free University of Bozen-Bolzano, Italy

Technical Program Chair

Pierpaolo Vittorini   University of L'Aquila, Italy

Paper Co-chair

Tania Di Mascio   University of L'Aquila, Italy
Fernando De la Prieta   University of Salamanca, Spain
Ricardo Silveira   Universidade Federal de Santa Catarina, Brazil
Marco Temperini   University of Roma, Italy

Proceedings Chair

Ana Belén Gil   University of Salamanca, Spain
Fernando De la Prieta   University of Salamanca, Spain

Publicity Chair

Demetrio Arturo Ovalle   National University of Colombia, Colombia
  Carranza

Workshop Chair

Elvira Popescu   University of Craiova, Romania

Local Organizing Committee

Juan Manuel Corchado Rodríguez	University of Salamanca, Spain, and AIR Institute, Spain
Fernando De la Prieta	University of Salamanca, Spain
Sara Rodríguez González	University of Salamanca, Spain
Sonsoles Pérez Gómez	University of Salamanca, Spain
Benjamín Arias Pérez	University of Salamanca, Spain
Javier Prieto Tejedor	University of Salamanca, Spain, and AIR Institute, Spain
Pablo Chamoso Santos	University of Salamanca, Spain
Amin Shokri Gazafroudi	University of Salamanca, Spain
Alfonso González Briones	University of Salamanca, Spain, and AIR Institute, Spain
José Antonio Castellanos	University of Salamanca, Spain
Yeray Mezquita Martín	University of Salamanca, Spain
Enrique Goyenechea	University of Salamanca, Spain
Javier J. Martín Limorti	University of Salamanca, Spain
Alberto Rivas Camacho	University of Salamanca, Spain
Ines Sitton Candanedo	University of Salamanca, Spain
Daniel López Sánchez	University of Salamanca, Spain
Elena Hernández Nieves	University of Salamanca, Spain
Beatriz Bellido	University of Salamanca, Spain
María Alonso	University of Salamanca, Spain
Diego Valdeolmillos	University of Salamanca, Spain, and AIR Institute, Spain
Roberto Casado Vara	University of Salamanca, Spain
Sergio Marquez	University of Salamanca, Spain
Guillermo Hernández González	University of Salamanca, Spain
Mehmet Ozturk	University of Salamanca, Spain
Luis Carlos Martínez de Iturrate	University of Salamanca, Spain, and AIR Institute, Spain
Ricardo S. Alonso Rincón	University of Salamanca, Spain
Javier Parra	University of Salamanca, Spain
Niloufar Shoeibi	University of Salamanca, Spain
Zakieh Alizadeh-Sani	University of Salamanca, Spain
Belén Pérez Lancho	University of Salamanca, Spain

Ana Belén Gil González   University of Salamanca, Spain
Ana De Luis Reboredo   University of Salamanca, Spain
Emilio Santiago Corchado   University of Salamanca, Spain
    Rodríguez
Angel Luis Sánchez Lázaro   University of Salamanca, Spain

Contents

Workshop on TEL in Nursing Education Programs (NURSING)

Workshop on Technology Enhanced Learning in Nursing Education

In the field of nursing, learning outcomes involve nurses both as learners and as educators.

As learners, they are involved in basic and post-basic academic programs, whereas they act as educators when they are engaged in health educational programs to enhance health literacy levels in the community.

In nursing research, learning outcomes have been widely investigated on different target populations, such as nursing students and staff, as well as patients and their caregivers.

However, the current exponential growth of the technology in the educational field makes it necessary to explore its contribution in enhancing expected outcomes for these populations, in order to facilitate the development of more accurate guidelines, protocols, and procedures.

According to some evidence, the quality of learning outcomes in basic and post-basic nursing academic programs could be potentially improved through technology-based systems that represent the basis for creating smart environments, where models like the high-fidelity simulation deserve great attention for the development prospects that they offer.

However, at this regard more robust confirmations are needed, as well as to discuss ethical and philosophical implications of technology enhanced learning in the field of caring. Furthermore, little is known about the use of technology to enhance health literacy levels in the community.

For these reasons, this workshop aims to share the best available knowledge about the application of technology-based systems into basic and post-basic nursing academic programs, as well as health educational programs aiming to enhance health literacy levels in the community.

In order to pursue this intent, workshop topics have been grouped into the following three main discussion points

First, topics on education in nursing academic programs aim to discuss the effects of simulation and other technology-based systems on learning quality, including ethical, legal, and philosophical perspectives.

Secondly, topics on community health educational programs aim to discuss the impact of technology in improving health literacy levels in the community.

Finally, the workshop intends to provide a complete overview of technology-based methods as useful tools to improve the learning of the nursing process in clinical settings.

The 2nd edition of MIS4TEL Workshop on Technology Enhanced Learning in Nursing Education includes ten accepted papers, significantly increased in comparison with the previous edition.

I would to thank the authors, the reviewers, Professor Rosaria Alvaro, my co-chairman, and the PC members Professor Elvira Popescu and Professor Pierpaolo Vittorini, whose support made this work possible.

Organization

Organizing Committee

Loreto Lancia   University of L'Aquila, Italy
Rosaria Alvaro   University of Tor Vergata, Italy
Elvira Popescu   University of Craiova, Romania

Program Committee

Angelo Dante   University of L'Aquila, Italy
Carmen La Cerra   University of L'Aquila, Italy
Cristina Petrucci   University of L'Aquila, Italy
Ercole Vellone   University of Tor Vergata, Italy
Pierpaolo Vittorini   University of L'Aquila, Italy

Problem Solving Incorporated into Blending Learning in Nursing Masters Degree

Loredana Pasquot$^{(\boxtimes)}$ ⓘ, Letteria Consolo ⓘ, and Maura Lusignani ⓘ

University of Milan, Carlo Pascal 36, 20133 Milan, Italy
{loredana.pasquot, letteria.consolo,
maura.lusignani}@unimi.it

Abstract. Online and face-to-face learning are integrated in a teaching format called blended learning. In recent years, educators have begun to use blended learning for a number of education related purposes. Typically, blended learning is used to involve the nurse students in a more active and constructive learning process. In a pilot project, five modules of a Masters nursing course were redesigned and implemented in blending learning format. While redesigning the modules, the first challenge was to assure the balance between online and face-to-face classroom activities. The second was to incorporate problem solving phases into blended learning in an efficient way. Moodle is the learning management system used for the online teaching and learning activities. The preliminary results concern the description of the redesign process of the five modules and their implementation. Some critical issues emerged and they must be corrected to improve the teachers' involvement and the redesign.

Keywords: Nursing education · Ill problem · Mind map · Problem solving · Blended learning

1 Introduction

Experiment of Enhanced Learning (EXEL) is the University's project of innovation in education (2017–2019). It is devoted to the promotion of innovative teaching and learning methods and an educational approach that combines classroom and distance learning.

The strategic aim of EXEL is to educate the teachers about student-centered teaching and learning rather than only being concentrated on the teacher and on the discipline, called teacher-centered approach [1]. This is done by the use of active teaching and learning methodologies and technologies.

The EXEL actions are planned in short, mid, and long period of time. In the short term, meetings with groups of teachers were organized to guide them in the teaching and learning innovation process. The innovation process is focused on the crash course "Arena blended curriculum ABC" (adapted to the Italian system with University College London collaboration) [2] and the online course "How to re/design your course with blended learning" (adapted to the Italian system with Utrecht University collaboration) [2].

© Springer Nature Switzerland AG 2020
E. Popescu et al. (Eds.): MIS4TEL 2019, AISC 1008, pp. 5–11, 2020.
https://doi.org/10.1007/978-3-030-23884-1_1

The Nursing Masters (second cycle, 120 European Credit Transfer and Accumulation System ECTS) program has been included in the EXEL project as a pilot project.

The redesign of the five modules was the aim of the pilot project for the Health and Prevention Health Care in the Community project. In this course, lot of the nurse students are already working, so the blended learning was seen as an effective way to facilitate both an active-constructive learning process and a redistribution of the teaching and learning hours.

The University's rule of the pilot project allows that only 10% of conventional hours in classroom can be used as online activities. So concretely, a module of 1 ECTS was divided into 2 h of online activities, 6 classroom activities and 17 autonomous study hours (1 ECTS = 25 h).

While redesigning the modules, the challenge was to assure the balance between online and classroom activities and the resonance between three components: learning design, student's learning experiences and the practice of the learning discipline. The redesign of the modules took place in the first six months of 2018, while the implementation of the modules began in November of the same year and ended in 2019, January 31.

Moodle is the learning management system used for the online teaching and learning activities of the students both in small groups and in autonomous study.

2   Tuning the Balance Between Online and Classroom Activities

Health and Prevention Health Care in the Community is the integrated course of the masters program that was redesigned to incorporate blended learning. It comprehends eight modules (one module = 1 ECTS), of which only five were redesigned, as you can see in Fig. 1. The redesign of the course did not include three modules because of three teacher's reluctance to replace their traditional forms of face-to-face teaching.

While studying the five redesigned modules, some elements effected the balance be- tween online and face-to-face learning activities. Of these elements the following were observed: more involvement of the students in the construction of their learning; more interaction in the classroom; more interest of students toward specific topics; deeper learning instead of memorizing; more capacity to analyze and synthesize health needs in the community. Based on the learning outcomes of the Health and Prevention Health Care in the Community course, the work started by focusing on the need to strengthen the coherence and the synergy between the five modules instead of the past method.

The coherence and the synergy between the five modules were considered essential components, in promoting in the students a mental habit aimed at seeking alternative solutions to prevention and care priority health needs of people at different stages of life and in continuous care.

Synchronic coherence required the tuning of the balance between online and face-to-face in classroom teaching and learning activities (how the learning on a number of separate, but synchronously taught, modules is experienced [3]). So the question was how to facilitate the students experience throughout the five modules? An online interactive activity was the solution. Concretely, in this online activity the students worked in small group to analyze three ill problems, as a first activity in each of the modules. Thirty minutes were removed from the total times scheduled for student's activities in each module (sixty minutes instead of ninety). This was done to give the students time to be able to do common online activities from the five modules (Table 1).

Another essential question was the sequential didactic coherence inside each individual module and the balance between online and classroom activities. In the case of doing online activities first, students came better prepared to the class. This is before the class the student had read documents watched videos uploaded online by the teacher about specific topics (Fig. 1).

Table 1. Architecture and timing of the five modules

Teaching and learning activities in common in the five modules (light blue space below)	Contents of activities	Online activities		Face – to- face class
		Teacher presents topics (Dispensing activity)	Student e- learning activity (Interactive activity)	
The five modules (white space below)				
First meeting	Presentation modules			60 m
Pre-test	Student's prior Knowledge		25m	
Analysis ill problems, mind map	e-Activities in small group		90m	
General clinic and pediatric nursing	Specific topics	30 m	60 m	6 h
Obstetrics and gynecology	Specific topics	30 m	60 m	6 h
Nursing and midwifery	Specific topics	30 m	60 m	6 h
Medical genetics	Specific topics	30 m	60 m	6 h
General and subspecialty pediatrics	Specific topics	30 m	60 m	6 h
Solution ill problems	Written Papers			

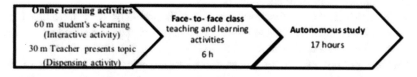

Fig. 1. Single module architecture and timing

3 Method and Theoretical Framework

The following learning outcome "Identify alternative solutions to priority health problems, along the life cycle, in prevention and continuous care" was the reference point for the redesigning of the five modules. This learning outcome embraces decision making, as an essential component in the nursing profession. In the modern health community environment, even with a high chronicity and disability level, nurses must be able to resolve problems of both the single patient and specific groups of people and try to do this effectively, efficiently and safely. So from a practical point of view nurses must make and justify decisions, reflecting on social and ethical responsibilities as well as nursing and nursing science issues and if necessary carry out analysis that results in an adequate basis for decision-making [4].

Nursing students need to receive training to foster strong problem solving and decision-making skills that can be utilized in the field [5]. Decision making and problem solving are two linked processes involved in the reasoning and the analysis of problems and the generation of solutions.

In the generation of solutions nurses often need to go beyond routine and acquire creative thinking to make beneficial decisions [6].

In Chan's systematic review, it is suggested that when designing a course to teach creative thinking, educators should provide activities, assignments or problem cases that allow students to use their creativity freely. Educators should value all ideas, encourage students to think, and give feedback that will guide students in the right direction. Group work and group interaction should also be considered as they may promote creative thinking [6].

Everything written above comprises the elements, which determined the choices of how to redesign the teaching and learning activities and their contents of the five modules incorporated into blended learning. Basically the problem solving method was adapted in a creative way, to seek the best way to marry the online activities with the need to grow the capacity to analyze problems and generate solutions by the students.

The principal components of problem solving incorporated into blended learning are: ill problems, mind maps, small group work, teacher's topic presentation and written solutions of ill problems. These components were organized into two steps of the method. In the first step, we define the problem framing by expanding our perspective [7]. A very significant part of this involves making sense of a complex situation in which the problem occurs, so we can pinpoint what the problem is. The second step, generating and evaluating alternatives, involves opening minds to see the problem solution from different point of view.

In the redesign of the five modules, three ill problems were introduced at the beginning as an initial learning situation. This approach helps the students to understand what they are learning in the five modules.

The first online activity for the students was an autonomous reading of the ill problems to identify some significant questions raised by them. The ill problems, in contrast to being well-structured, may have many possible answers and are complex and poorly defined. Ill-structured problems, require the development of higher order thinking skills and the ability to construct a convincing argument for a particular solution as opposed to all other possible solutions. So, for the problem to be effective for the purpose it must generate perplexity, confusion or doubt [8].

In Moodle platform a shared environment was opened and there the students found and read the three ill problems uploaded by the teachers.

The second online learning situation was designed for a small group context aimed at analyzing the problems and at constructing two mind maps for each of the three ill problems. The students were divided into five randomized groups where to work collaboratively to shape, elaborate and deepen understanding [9].

Each of the small groups worked on the platform Moodle with Wiki to share the analysis of ill problems, and write down their questions and ideas related to each of the questions. The picture of this Wiki's environment shows the result of a virtual brainstorming session inside the small group.

A Chat was opened in Moodle, where the small groups, in asynchronous discussion, compared their questions and ideas and chose those more significant in the definition of ill problems. After that, each of the small groups synthetized its decision in two mind maps for each of the three ill problems. Due to the lack of the mental mapping tool in Moodle, students were asked to use Power Point. The mind maps built by Power Point were sent to the teachers through the Wiki's section File.

This approach to the collaborative group, shows that analyzing ill problems and synthesizing priority question and ideas in mind maps, is consistent with the idea that work- ing together allows the students to compare alternative interpretations, correct each other's wrong ideas and form more holistic pictures of the problem. According to some authors, effective learning, is collaborative and social rather than isolated, and working with others increases involvement in the learning process [10].

The mental maps were used by the students to organize ideas around a central topic question, visualizing associations and deductions through cross-links. The mind maps are principally association maps, and spontaneous thinking is required when creating them [11].

4   Teaching and Learning Online and Face-to-Face Activities in Each Module

Each of the five modules was redesigned by the teacher as showed in Fig. 1 (one teacher in each module). The time for online activities inside the module are fixed at sixty minutes for e-learning interactive activities. Then there is also thirty online minutes for the online dispensing activities (the material uploaded by the teacher can last a maximum of 15 min, so the student can look at the material at least twice within the 30 m provided for the online dispensing activity). On the Moodle platform there was an autonomous environment for each of the modules, where the students could find the material uploaded by the teacher. A Moodle board was also opened for interactions and quick feedbacks.

The sequential coherence was studied by the teachers, aligning the learning activities with specific learning outcomes and monitored the students online and face-to-face activities. The decision to do the students online activities before the face-to-face class activities, was aimed at having students come better prepared to class. In this way the class became a place of discussion under the direct supervision of the teacher.

5   Conclusion

Unfortunately, as the implementation of the five modules ended in 2019 January 31, the results of students feedback are not available in this presentation. Therefore the presentation was focused on the redesign of the five modules of the Health and Prevention Health Care in the Community Course, which were described in their didactic components and how they incorporated blended learning. However, some reflections about critical issues are possible to make.

Unstructured interviews were conducted with the teachers of the five modules, to deepen their perceptions of the various components of problem solving incorporated into blended learning.

Teachers perceived an increase in their workload because of the need to find, select and upload the online material that students had to learn before the classroom activities (15 min in which the teachers dispensed the material to the students). This additional work was done to prepare for the face-to-face lessons in the classroom. The teachers are aware, however, that this increase in workload will not occur in subsequent applications of the redesign, because it will only be necessary to replace or update the material already loaded.

Some teachers also point out the difficulty of finding educational published material in Italian, such as videos, suitable for students of the nursing masters degree. This difficulty has been perceived as an obstacle to an effective use of online learning and requires the preparation of ad hoc material to be overcome.

Moodle technology has been a novelty for teachers and the various functions available have not been used to the full. This is because a competent use of Moodle would have required a longer period of training than the teachers actually had. That being said the teachers, however, have positively evaluated the technological support received for the uploading of the material online, but they complain about the need to acquire greater autonomy.

The limited autonomous use of Moodle technology by the teachers and the limited time available to learn how to use it has caused a very poor realization of the interactive online activities (only one teacher has been able to use this online interactive activity for discussing a clinical case). Teachers consider it important to integrate their modules with these activities, in particular with case discussion groups and exercises, to make the best use of the 60 min available to them for interactive activities.

The problem solving method used as a common activity in the five modules has been positively evaluated by the teachers. They believe that the mind maps produced by students, working online in small groups, are useful in getting to know the students' ideas regarding the three ill problems. According to the teachers, understanding these ideas helps them in their approach to the teaching of the contents of the modules and thus facilitate the students in finding solutions to the three ill problems. However, the teachers consider it necessary to improve the context of the problem framing, making the best use of technological resources to manage the online activities in small groups.

A conference call environment is needed to enable the students to collaborate effectively in problem analysis and mental map processing. In this conference call, the presence of a tutor (even the teacher themselves) should be guaranteed to assist the students and, if necessary, to intervene, to urge, and to restate the problem in different ways, and to select those ones best suited for exploration. In addition, more specific tools for building mind maps should be integrated into Moodle, to visually organize what is known about the problem.

The results of the interviews reveal some critical points but at the same time hypothesize solutions that will be used to correct the application of the redesign of the five modules and create a more effective strategy of problem solving incorporated into blended learning. Feedback will also be conducted with the students, whose insight will be used to discover other ways of improving education through online resources. The online resources in blended learning are a very important aspect as it is more adapted to the educational needs and lifestyle of students, especially working-student.

References

1. Kennedy, D.: Writing and Using Learning Outcomes: A Practical Guide. University College Cork, Cork (2006)
2. Piano Strategico di Ateneo 2017/2019. Università degli Studi di Milano. http://www.unimi.it/cataloghi/comunicazioni/Piano%20strategico%20di%20Ate-neo%202017-19_%20Universita%20degli%20Studi%20di%20Milano.pdf
3. Carnell, B., Fung, D.: Developing the higher education curriculum, pp. 21–22. UCL PRESS, London (2017). https://doi.org/10.14324/111.9781787350878
4. Tuning Guidelines and Reference Points for the Design and Delivery of Degree Programmes in Nursing 2018 edn. (Agreement number 2015-2666/001-001)
5. Heidari, M., Shahabazi, S.: Effect of training problem-solving skill on decision-making and critical thinking of personnel at medical emergencies. Int. J. Crit. Illn. Inj. Sci. **6**(4), 182–187 (2016). https://doi.org/10.4103/2229-5151.195445
6. Chan, Z.C.Y.: A systematic review of creative thinking/creativity in nursing education. Nurse Educ. Today **33**(11) (2012). http://dx.doi.org/10.1016/j.nedt.2012.09.005
7. Vernon, D., Hocking, I., Tyler, T.C.: An evidence-based review of creative problem solving tools: a practitioner's resource. Hum. Resour. Dev. Rev. **15**(2), 230–253 (2016). https://doi.org/10.1177/1534484316641512
8. Yew, E.H., Goh, K.: Problem-based learning: an overview of its process and impact on learning. Health Profession Educ. **2**(2), 75–79 (2016). https://doi.org/10.1016/j.hpe.2016.01.004
9. Biggs, J., Tang, C.: Teaching for Quality Learning at University, 4th edn. MacGraw-Hill, New York (2011)
10. Schwartz, B.M., Gurung, R.A.: Evidence-based teaching for higher education. Am. Psychol. Assoc. (2012). https://doi.org/10.1037/13745-000
11. Davies, M.: Concept mapping, mind mapping and argument mapping: what are the differences and do they matter? Higher Educ. **62**(3), 279–301 (2011). https://doi.org/10.1007/s10734-010-9387-6

Nurse Application Bundle: Tools for Nursing Practice in Home Care Settings

Francesco Biagini[✉], Scialò Gennaro, Macale Loreana,
Palombo Antonella, Vellone Ercole, and Alvaro Rosaria

Biomedicine and Prevention Department, University of Rome "Tor Vergata",
via Montpellier 1, 00133 Rome, Italy
biaginifrancesco93@gmail.com,
gennaro.scialo@gmail.com, loreana.macale@gmail.com,
palomboanto@aol.com, ercole.vellone@uniroma2.it,
rosaria.alvaro@gmail.com

Abstract. New technologies are becoming relevant for nursing practice in the home care field. Specifically, Mobile Applications can be useful tools for nurses to care for patients efficiently and to ensure optimal data protection with the help of electronic folders (HER) and with utilities that can be used as aids for work analysis in care planning.

Keywords: Mobile Application · Nursing homecare · House assistance · Technology · Digital health

1 Introduction

The word we have known in the last century does not exist anymore, Welfare models need to be revised [1]. As a matter of fact we are witnessing a crisis in the healthcare systems mostly triggered by the demographic change of the population [2]. It is estimated that by 2050 there will be 1.6 billion elderly people around the world for a total of 16.7% of the world population of 9.4 billion [3]. This condition is leading to a great increase of chronic disease [2] which, can also cause disability [4] such as reduction or loss of functional capacity [5]; with an increase of weakness [6]. Therefore, awareness and demand for healthcare are increasing among citizens [2]. For this reason, countries need to implement innovative ideas and models that take into account the real problems of the population, with the purpose to promote solutions to allow Welfare systems to be able to adapt and transform without collapsing [1].

Therefore, it is necessary to rethink the concept "health equal hospital" by shifting the attention from the hospital to the territory and by looking at home care as the best possible answer [7]. International experiences regarding chronic illnesses' management in the home care setting show that nurses play a very important role [8] as they bring their experience and expertise to improve health. Nurses who practice in home care settings have good critical thinking and teaching skills, are good listeners and have great communication with patients' families and caregivers [9].

Parallel to home care, it is also important to look at technological progress for patients' benefits [2] and for clinical practice in the home setting itself [10] of new

E. Popescu et al. (Eds.): MIS4TEL 2019, AISC 1008, pp. 12–17, 2020.
https://doi.org/10.1007/978-3-030-23884-1_2

technologies such as Information and Communications Technology (ICT), systems useful for electronic health registration (HER); portals for patients; eHealth; Mobile Applications [11], managed by expert nurses who supervise and use the new technologies and who have assumed, and will assume more and more, a role of leadership [12]. The advantages offered by Information Technology (IT) are: a better quality of care for people; offer greater connection with the rest of the field staff; improve communication; be able to get more work done and reduce costs. In the home care field, organizations are taking ad advantage of new technologies such as monitoring devices for patients and technology that allows the integration of electronic medical records (HER). Each of the new technologies foresees the need of data protection to ensure patients' privacy [13]. Home care nurses can then adopt new technology tools [14], with the possibility to daily update documentation by reducing the use of paper and increasing its level of accuracy. All of these tasks are possible thanks to the continuous updating that takes place immediately, such as the possibility to view the patient's files at any time throughout the shift, plan a work schedule and safeguard the health of the client [15]. In the past the concept of good documentation was often associated with a great amount of recorded data which would have improved care, but today nurses believe that documentation can have just few but relevant information. It must be clear, rational, objective, comprehensible and has to accurately reflect the patient's health conditions and the obtained outcomes [16].

Therefore, it is real the need to have complete nurses' charting for continuity of care in home care settings [17] and with technological progress it is possible to develop standardized care plans to ensure a logical sequence of nursing interventions [14].

Electronic medical records (HER) facilitated data recording and improved the nursing process [18]. Likewise, Mobile Applications are evolving and increasingly require nurses' attention [19] as they can be incorporated into the electronic health documentation and can be considered suitable, robust and well designed to pursue the needs of the patient [12]. Nurses are considered fundamental to connect health to the health care of the future [11] throughout e-health which is a suitable solution [20]. Specifically, Mobile Applications are programs or tools that are used by information and communication devices (ITC) such as smartphone and tablets, in order to perform specific functions. Mobile Applications in the heath field can be categorized into three different domains: (1) Applications used by consumer, i.e. patients; (2) Application used by health professionals; (3) Useful applications to communicate/or control a device that interfaces with other ICT [18]. Nowadays, the use of mobile aids such as smartphones and tablets at a territorial level plays a predominant role [21]. In literature, there are examples of Mobile Applications for home care such as the one regarding patients with colostomy [22]. It is also possible to download nursing applications on the Play Store platforms such as the Lippincott Nursing Advisor which is an Applications (App) where a nurse answers to all kinds of questions regarding diseases, sing and symptoms, exams and treatments [23]. Therefore, the prospective scenario is that of rapid growth and broad sharing which is leading to the improvement of healthcare outcomes [24].

1.1 Aim

The aim of this project includes: (i) improvement of the fundamental process of care, (ii) improvement of the knowledge regarding the nursing process, and (iii) Promote the production of clear and quality documentation.

2 Nurse Application Bundle

The Nurse Application Bundle is a useful Mobile Application for nurses, create by Francesco Biagini, one of the authors of this manuscript, with the aim of improving clinical practice quality. The Mobile Application is compatible for both smartphone and tablet with an Android system ranging from 4.4 Kit-Kat to 8.0 Oreo. Specifically, the Nurse Application Bundle contains Information of technologic Communication (ICT), data collection and registration of the electronic folder (EHR), all reported on a native App. Also, an interference link between the App and database for saving, storing and securing the data of the patient was created. The chosen and interfaced database is a Google database made available by the "Freebase" package, a no-sql, which guarantees a complete compatibility with different kinds of drivers. The App foresees a mandatory registration with personal credentials: username and password. The App presents a main home screen with 3 sections: (1) Add user; (2) User folder and (3) Utility. Each one of these sections allows the nurse to perform different actions. The App user section contains the electronic medical record (EHR), which allows to produce a care plan for each patient. The care plan is essentially divided into different parts to keep track of the chronological process to guarantee efficient and quality care. The points contained in it are: Add new user (this part is useful to register each patient to system and create their own space dedicated to documentation); Personal data; Vital signs; Medical history; Assessment; Evaluation scales; Targets; Interventions and Final Evaluation. Nurses will be have these information displayed in digital format on their devices and will be able to fill them out in touch-screen mode in the appropriate designed fields. The user folder section, on the other hand, allows viewing of patients' documentation that has been expressed on different days and saved by the nurse. The monitoring and changes over time of the documentation are subdivided by days, each access is viewable and represent an update of the documentation and this also implies an optimal traceability of the provided care. All data collected and present on this section, as already mentioned above, are saved in the database that is interfaced to the App. Finally, the utility section, allows the nurse to have various tools: agenda, price list; medication calculation; procedures; good practice; camera. The agenda is useful for nurses to make appointments and organize daily work schedule. The price list contains the national price list related to nursing services. In this section, nurses can use the nomenclature and enter their payment for each professional service. Also, it is possible to create a personal price list, nomenclature and compensation [24]. The medication calculation section includes several calculators: a unit of measure converter; a calculator for the dosage; a millimeter-hourly calculator and a milliliter-per-minute calculator related to the rate of intravenous flow to be infused. The procedures section allows the nurse to consult technical procedures if necessary. The procedures have been extrapolated from the reference the Kozier and Erb's textbook "Techniques in Clinical Nursing" [24]. The best practices can also be consulted in the temporary absence of the new guidelines, as these are being re-elaborated, as established by the Law n. 4 from March 8, 2017 (the so called Gelli Law) [25]. The best practices have been extrapolated from the various official websites of Nurses Associations but also from other health associations, namely: Nurses Association of Urology (AIURO); National Nursing Association of Critical Care Nurses (ANIARTI); National Nursing Association of Hospital Medicine (ANIMO); Scientific Society of Specialist Nurses in Infectious Rehabilitation (ANIPO); Italian Palliative Care Society (SICP); Association of Diabetes Doctors (AMD) and National Federation of Nursing Professions (FNOPI) [27]. Finally, the camera section, is useful to the over time monitoring of situations that require special treatment. In order to guarantee patients' privacy the nurse must collect patients' consent and note it in the interventions before taking any pictures.

3   Direction for Nursing Practice, Research, and Learning

The Nurse Application Bundle is easy to use thanks to its simple design. The App will allow home care nurses to organize their work; manage the workload of a large number of patients; reduce the amount of paper used for charting. Also, the instrument cloud be consulted during work, thanks to the tool; create and update patients' document in a fast and sequential manner and finally be able to provide patients and/or their family with a chart of the received nursing care available on a digital support saved file or on cloud. However, the Nurse Application Bundle can be a valid instrument for learning because every part of the patient's chart format contains all the fundamental steps for care planning and by using the Application, nurses have the chance to improve their knowledge about the nursing process day after day by producing clear and high quality documentation. Also, there are other aspects of the Application, such as the Evaluation scale part, which are very useful for learning: in this case, nurses can look at the user's instructions under each scale where they can find all the information with the purpose of nursing the scale appropriately. Because of its detailed characteristics regarding the nursing process, nursing procedures and medication dosage calculation, the Nurse Application Bundle could also be used by nursing students to learn and to develop the critical thinking skills necessary for their future practice. This App is unique in the home health care setting because with the use of multiple users care outcomes could be easily assessed to understand the degree of usefulness for the Application and the real ability of the database to be able to contain a vast number of data stored and saved in it. One of the tool's limitation is that, as the project will expand and more storage will be required to memorize and save patients' data, it may be necessary to implement a dedicated server with full backend functionality with a no-sql MongoDB database.

References

1. Maino, F., Ferrera, M.: Terzo Rapporto sul secondo welfare in Italia (2017)
2. Value in Health Technology and Academy for Leadership and Innovation (VIHTALI). L'evoluzione delle modalità di finanziamento dei sistemi sanitari nazionali. Roma (2017)
3. He, W., Goodkind, D., Kowal, P.: An aging world: 2015. International population reports. P95/161 (2016)
4. Petersen, P.E., Ogawa, H.: The global burden of periodontal disease towards integration with chronic disease: prevention and control. Periodontology 2000 **60**(1), 15–39 (2012)
5. Nebuloni, G.: Pianificare l'assistenza agli anziani nel ventunesimo secolo. Casa Editrice Ambrosiana, Milano (2012)
6. U.S. Department of Health & Human Service: "What is Long-Term Care?" (2017)
7. Vetrano, D.L., Vaccaro, K.: La Babele dell'assistenza domiciliare in Italia: chi la fa, come si fa. Indagine, Roma (2018)
8. Marcadelli, S., Obbia, P., Prandi, C.: Assistenza domiciliare e cure primarie. Casa Editrice Edra, Milano (2018)
9. Elliott, B.: Considering home health care nursing? Nursing 2018 **44**(12), 57–59 (2014)
10. Tibaldi, V., Ricauda, N.A., Rocco, M., Bertone, P., Fanton, G., Isaia, G.: L'innovazione tecnologica e l'ospedalizzazione a domicilio. Recenti Progressi in Medicina **104**(5), 181–188 (2013)
11. Murphy, J., Honey, M., Newbold, S., Weber, P., Wu, Y.H.: forecasting informatics competencies for nurses in the future of connected health. Stud. Health Technol. Inf. **250**, 58–59 (2018)
12. Rojas, C.L., Seckman, C.A.: The informatics nurse specialist role in electronic health record usability evaluation. CIN: Comput. Inf. Nursing **32**(5), 214–220 (2014)
13. Grande, E.: Information technology and home healthcare: the new frontier in home care. Home Healthc. Now **32**(3), 194–195 (2014)
14. Schnelle, J.F., Bates Jensen, B.M., Chu, L., Simmons, S.F.: Accuracy of nursing home medical record information about care process delivery: implications for staff management and improvement. J. Am. Geriatr. Soc. **52**(8), 1378–1383 (2004)
15. Daniels, R., Grendel, R.N., Wilkins, F.R.: Basi dell'assistenza infermieristica. Casa Editrice PICCIN, Milano (2014)
16. Carpenito-Moyet, L.J., Calamandrei, C., Rasero, L.: Piani di assistenza infermieristica e documentazione: diagnosi infermieristiche e problemi collaborativi, 2nd edn. CEA Casa Editrice Ambrosiana, Milano (2011)
17. Gjevjon, E.R., Hellesø, R.: The quality of home care nurses' documentation in new electronic patient records. J. Clin. Nurs. **19**(12), 100–108 (2010)
18. Schachner, M.B., Sommer, J.A., González, Z.A., Luna, D.R., Benítez, S.E.: Evaluating the feasibility of using mobile devices for nurse documentation. In: Nursing Informatics, vol. 225, pp. 495–499 (2016)
19. Elias, B.L., Fogger, S.A., McGuinness, T.M., D'Alessandro, K.R.: Mobile apps for psychiatric nurses. J. Psychosoc. Nurs. Ment. Health Serv. **52**(4), 42–47 (2013)
20. Granja, C., Janssen, W., Johansen, M.A.: Factors determining the success and failure of e-health interventions: systematic review of the literature. J. Med. Internet Res. **20**(5) (2018)
21. Associazione Italiana Sistemi Informativi in Sanità, Mobile Health (2014). http://www.aisis.it/it/workgroup/gruppo-di-lavoro-2014-mobile-health/97d81d74-8d45-49b682c4-0332dc54ec7a

22. Wang, Q.Q., Zhao, J., Huo, X.R., Wu, L., Yang, L.F., Li, J.Y., Wang, J.: Effects of a home care mobile app on the outcomes of discharged patients with a stoma: a randomized controlled trial. J. Clin. Nursing **27**, 3592–3602 (2018)
23. Play Store: Lippincott Nursing Advisor. https://play.google.com/store/apps/details?id=com.Iww.Ina.mobile.android
24. Marucci, A.R., De, W.C., Petrucci, C., Lancia, L., Sansoni, J.: ICNP-international classification of nursing practice: origin, structure and development. Professioni infermieristiche **68**(2), 131–140 (2015)
25. Berman, A., Snyder, S., Alvaro, R., Jackson, C., Brancato, T.: Nursing clinico. Tecniche e procedure di Kozier, 2nd edn. Napoli, Italia (2012)
26. Gazzetta Ufficiale della Repubblica italiana. www.gazzettaufficiale.it/eli/id/2017/03/17/17G00041/sg
27. Federazione Nazionale Ordini Professioni Infermieristiche. http://www.fnopi.it/

Care Robots: From Tools of Care to Training Opportunities. Moral Considerations

Maurizio Balistreri[1](✉) and Francesco Casile[2]

[1] Department of Philosophy and Educational Sciences, University of Turin,
Turin, Italy
maurizio.balistreri@unito.it
[2] Bioethics Committee, University of Turin, Turin, Italy

Abstract. New technology should not perceived as a threat: on the contrary, it
is a resource. It is our job to use it in the most appropriate way. Moreover, new
technology can make a significant contribution to nurse training: for example,
immersion into virtual reality with a visor and a simple application does not only
allow one to experience fantastic adventures, but also to enjoy a relationship
with the patient through simulation. Also, virtual reality can promote
patient/teacher interaction: both, for example, can be projected or immersed in
virtual reality, or the teacher can project his 'virtual' image into a real scenario.
However, robots too could contribute to training nursing staff: health operator
training courses today widely use dummies which are appropriately planned for
standing training. They are increasingly true-to-life, favouring empathy with the
clinical situation simulated each time and allowing the student to exercise not
only technical abilities, but also critical thinking, the ability to work in a team
and communication skills. We shall examine some moral questions linked to the
increasingly frequent use of human-faceted robots to train nursing staff.

Keywords: Carebots · Robots · Bioethics

1 Robots in Caring and Healing Roles

Over the last few decades, the use of robots has markedly increased and it is easy to
predict that they will be even more a part of our existence in the decades to come.
Robots also promise to be a significant resource in the field of medicine: especially
patients may derive advantages from their introduction and use, in that they may be
assisted by machines that are not subject to tiredness or stress, hence may be operative
24/7 without a break. Further, robots may not only monitor their health condition more
precisely than a health operator or human (for example, robots could be programmed to
note a significant change in the behaviour, habit or voice of the person followed), but
also compensate for or limit the impact of some physical disabilities. They could,
furthermore, assist a sick or elderly person just as well as a human: for example, there
are robots that can be programmed to remind the patient when it is time to take
medicine and which pharmaceuticals out of many to take, whilst others have the task of
stimulating their cognitive abilities through games or questions. In this way, they can
function as a both physical and cognitive prosthesis, in that a person who, for example,

E. Popescu et al. (Eds.): MIS4TEL 2019, AISC 1008, pp. 18–25, 2020.
https://doi.org/10.1007/978-3-030-23884-1_3

forgets to eat will be on the road to ruin in many ways, not least cognitively. And this type of thing may become a vicious circle [1]: the person continues not to eat, lucidity worsens, he forgets to eat again or take care of himself [2].

However; the use of robots in the field of care may also have important repercussions for health operators, as well as for anyone beside a fragile person, in that caregivers could have at their disposal a very useful tool [1] to use for activities that are more repetitive (for example, bringing medicines to the patient's bed, handing out breakfasts, lunches and dinner) and tiring (for example, lifting a patient from bed or helping him walk and exercise or work) [3]. In these terms, carebots could help people carry out their care job without getting ill or still having to take on unbearable burdens that are incompatible with a satisfying life with their family or those held dear: "caregiving practice" – writes Shannon Vallor – "in the absence of adequate resources and support have a highly destructive impact on emotional, prudent and social capacities – exasperated stress, anxiety, physical and exhaustion can cause emotional withdrawal, a gradual numbing of affection or extreme emotional volatility – one or all of these consequences can cause fractured relationships with those assisted and others, serious depression, motivational apathy, compromise ability of judgement and even lead to a total nervous breakdown [3]." It is true that the most well-to-do have always been able to choose to delegate part of their care work to their family or those they are responsible for (not only the elderly, but also children and individuals displaying disabilities) to others (not only care institution staff, but also caregivers or freelance workers) who see their work force and skills in care for others. However, the spread of robots could benefit the whole population, in that their cost could notably lower over time. Further, any initial expense for their purchase could be amortised over a period, in that they do not have to pay the robot a wage or overtime. And once purchased, the robot can be resold or possibly shared with others who need care and assistance. For this reason, robot technology could contribute to promoting social justice [1]", in that the care work would no longer be an activity imposed by circumstance, but the result of a freer choice that could be shared with those most interested.

Of course, spreading robots in the field of care also arouses worry. For example, there is the fear that with the use of increasingly intelligent robots, patients could be totally abandoned to assistance from machines; and at that point, human beings would – maybe forever – lose empathy, in that they would no longer get to experience others' suffering. However, we cannot imagine that robots, once introduced into the field of care, would substitute the health staff thus far employed fully and in all services. Robots may reveal themselves as able to replace health operators in those contexts requiring more simple and repetitive services (for example, supplying meals or medicine to patients at times that could be programmed), but not in those care activities needing to respond to more complex situations (not only diagnosis of illness and choice of type of operation or therapeutic treatment, but also the psychological assistance of the patient or elderly person). So the robots could provide a "superficial" or mechanical service [4] – for example performing repetitive manual tasks –, while the health operators (doctors and nurses) "deeper" care for the patient's specific individuality and needs: "While we have no reason to think that robots will be unable to offer individuals superficial, we share Coeckelbergh's scepticism regarding the possibility that they may go beyond superficial care and offer that type of reciprocal company or friendship that characterises deep care. (…) Robots' inability to offer deep care does not hinder human caregivers from using them to offer good care, respecting dignity [1]". Further, introducing carebots into medicine could open up to assistance and care new possibilities that we still struggle to imagine and which produce not the disappearance but the redefinition of care practices (least of all those we perform to meet the needs of others) [4]. For example, can we think that thanks to the very help of robots, elderly or ill people could count on better quality relationships with their family? People responsible for the care of a sick or elderly person (it does not matter whether they are family or paid by the family or an institution) can sometimes suffer from depression and discouragement or be overwhelmed by work and so not have the time or energy to dedicate to entertainment (to give the person they care for a pleasant, fun time). With robots programmed to carry out the more tiring, boring care work, things could change, in that family members or in any case others could use part of their time in relating to people who are not self-sufficient. It is also possible that part of the care activity would consist of laboratory planning of machines to deal with care and assistance. Indeed, one can imagine that robots should be personalised to take into account the needs and preferences of those they will then care for (after all, different patients will not only have different needs in terms of therapy, but also different habits, which the robot will have to take into account). It will therefore be the family members' task – as well as the health operators', of course – to rebuild the profile of the person and indicate the tasks and services the robot will have to perform. Of course, people who are still competent may choose with greater freedom the type of robot they want and the type of activity they can expect. However, the interests of those who are incompetent – unless perhaps they have left a trail of their own volition – must be safeguarded by others.

2 Interactive Robots for Training and Refining Clinical Skills

New technology should not be perceived as a threat: on the contrary, it is an important resource. It is up to us to use it in the most appropriate way. Carebots are more and more present by the patient's bed and can collaborate in care work. However, new technology can make a significant contribution to nurse training: for example, immersion into virtual reality with a visor and a simple application does not only allow one to experience fantastic adventures, but also to enjoy a relationship with the patient through simulation. Also, virtual reality [5, 6] can promote patient/teacher interaction: both, for example, can be projected or immersed in virtual reality, or the teacher can project his 'virtual' imagine into a real scenario and assist health operators. However, robots too could contribute to training nurses and give them the chance to practice clinical skills before they see a real patient. Health operator training courses today widely use dummies which are appropriately planned for standard training [7–9]. The models' physical and anatomical features are increasingly true-to-life, favouring empathy with the clinical situation simulated each time and allowing the student to exercise not only technical abilities, but also critical thinking, the ability to work in a team, and appropriate communication skills. One can imagine that increasingly intelligent and interactive robots could make training even more useful for refining clinical skills. Robots to train have already been constructed so as to let doctors and training nurses carry out plenty of tests, like measuring blood pressure or checking pulses and breathing [10]. Further, it can be much easier for teachers to check the student's level, in that the automaton is connected to a computer or able to register its results in various simulation tests: a "robot can be programmed to simulate the required actions in a way that can be objectively and quantitatively measured [7]." Also, according to didactic needs, the robots could be programmed to interpret people of different ages and different character profile: for example, the robot Hal, created by Gaumard, not only looks like a child, but also behaves like one: tracking a finger by eye is not his sole ability; he can also reply to questions, cry for his mother and feel anaphylactic shock. In this way, the student could face up to different situations and practice in working in pediatric conditions where the patient's compliance is partially or completely absent.

However, robots are not only useful in practising the student's technical skills: especially in basic training paths, robots could also be used to improve health operators' readiness to understand the patient's feelings more precisely (empathy). We are now used to 'mechanisms' which present themselves with sentiments and needs: which do not want to play with riddles, but say if they are hungry or sad and must be nourished, entertained and cleaned because otherwise they would die or not grow. It does not seem difficult to project a robot able to simulate the behaviour of a patient and with different reactions according to the way in which it is treated: for example, if the operator's care is adequate, it could smile or say thank you; but if it is neglected, it could cry or get angry and sad and even say, "You are mistreating me." To programme it, it would just be necessary to gather information relating to the quality or enjoyment of care, obtained by the patients via questionnaires or interviews: in this way, we will get to know their expectations regarding the health operators, what is important in them for care and what is, on the other hand, neglected. Of course we already have an abundant literature regarding the most appropriate behaviour towards a patient: but perhaps, if we want to build a robot able to simulate the behaviour of a patient, it will be necessary to develop thus far unconsidered aspects. Through interaction with robots, the health operator would first and foremost have a chance to evaluate immediately the appropriateness of its care, with no risk of self-deception or not adequately grasping the patient's reactions. How many patients are brave enough to complain about a lack of care? Moreover, children may not be forthcoming about their symptoms: they can often do so by facial expression or behaviour and gestures. While students carry out treatment, the robot could react to the prick of a needle by crying or expressing displeasure. He might also move, not only showing emotions and sentiments, but mimicking the difficulty often found when treating a person in pain. Further, due to illness, some people are unable to react to what is going on around them, and seem unable to have feelings: on the contrary, the robots will not feel any awe towards us or have any embarrassment and shame, in that – in any situation – they will do what they were programmed to display. Overall, we can self-deceive with people, but with robots this is more difficult. For this reason, health operators inter-reacting with robots can learn to calibrate their behaviour or relationship with the patient, and then modify it in the face of negative responses: once the robot starts to seem content and is no longer agitated, crying or complaining, then here we are: we have reached the adequate care level.

We can add to this that training with robots could make it easier for health operators to accept robots for care and assistance, and to recognise that robots are a resource that can not only lighten health operators' work but also improve the quality of care. We are actually at the start of a revolution in the field of training that could radically alter training nursing staff and doctors. Of course, we have not yet fully assessed the effectiveness of using robotics in training, but we do have cause for optimism [11–14]. Studies demonstrate that undergraduate student nurses can substitute as much as half of their clinical hours by simulations without affecting their ability to pass their nursing exam and complete their study path with flying colours. In fact, new research indicates that up to 50% of clinical hours can be replaced with simulation training [15, 16].

3   Robots for Training and Moral Issues

Over the last decades, many people have grown affectionate and constantly taken care of electronic devices as elementary as, for example, Tamagotchi: one can imagine that the more robots resemble human beings, the more we will worry for their wellbeing and it will be far easier to 'empathise' with them. Despite being 'corporeal' enough for a child to imagine its death, Tamagotchi is still a virtual creature in a plastic egg, whilst robots are machines that are increasingly taken for human beings. In these terms, interaction with these robots raises the same questions that emerge in relation to the use of robots in the field of care and affectionate relationships.

We do not yet know if our character can be affected by interaction with a robot: for example, there is the worry that, over time, violent behaviour toward a machine may turn one towards violence. Think, for example, of sex robots: could the use of sex robots programmed to refuse sexual relationships increase violence against women? And could someone spending a lot of time before a computer killing, raping and torturing other 'people' be then more likely to commit such crimes in real life? This is a complex question without scientific agreement: the possibility that care towards robots could promote empathy is only a marginal research theme, but it deserves greater attention, because the risk is that one starts to feel affection and tenderness towards a robot. For some people, the experience with the robot could turn out to be a disappointment, whilst for others interaction with the robot could seem an authentic relationship. For example, this is why Gaumard did not want to make its robot Hal too realistic and like a flesh-and-blood child. They feared such a robot that could also bleed to death or experience cardiac arrest might be too emotionally traumatic for student just being introduced to emergency pediatric situations". That is, in the field of training, robots are useful to let students implement their knowledge and learn from their mistakes. However, with overly realistic model students could get into difficulty, in that they could be afraid of erring and inflicting 'suffering' on the robot. Further, an excessively realistic android could distract participants during a tense scenario: their attention could be taken by irrelevant model details or they could carry out certain actions only to see the reaction caused. Moreover, if the operator interacting with the robot spends a lot of time with the machine or gets close to it, because he intends to care for it [17], this could negatively affect family relationships: a partner may be jealous; a child or friend could feel neglected. Virtual entities like Tamagotchi or LovePlus "girls" are not "real", but those who interact with them daily succumb to the same appeal and infatuation that children usually feel towards their favourite toys.

Moreover, the same phenomenon has been observed studying the use of robots in the field of caring for and assisting the elderly. These people are aware that the robot is not a human being, but for them the robot is unlike a mere machine and they think they have responsibility towards it. Also, people sometimes grow so fond of a doll that they end up loving it: a lot of imagination is not required to imagine what could happen tomorrow with robots that are increasingly intelligent, technological and maybe indistinguishable from ourselves. If the robot can produce the same levels of experience and company as a human, we may not resist the temptation to treat them like authentic human beings.

But the android could also arouse completely different reactions in the human being: it is the Uncanny Valley problem [18–21]. Indeed, it has been observed that the more human the robot seem and the more similarly they behave – but not in exactly the same way – to human beings, the easier it is to feel a great sense of revulsion and disgust towards them. A possible explanation for the phenomenon, first documented by Masahiro Mori [22], is that it would mean an adaptive reaction selected by evolution as useful to keeping our distance from ill or strange people (after all, the same repulsion is felt towards cadavers and zombies), but another hypothesis has been put forward too: the humanoid robot's unnatural movements would arouse thoughts of death in us [23]. Of course, this problem could make the use of robots as training tools for nurses in basic courses or even later more difficult. The simplest solution would seem to be making the robots less similar to human beings, but in this way students would probably have less chance to test and exercise their technical abilities. However, not only has Mori's theory been inadequately studied so far, but the repulsion and disturbance observed in the Uncanny Valley phenomenon could disappear with robots completely undistinguishable from human beings. Indeed, "a 'person in health' (or 'someone' perceived as a person in health) arouses a feeling of full acceptance, empathy and familiarity in the perception of another human being. A robot undistinguishable from a human being or person in perfect health does not yet exist, but will exist in the relatively near future [24]."

4 Conclusions

With the development of artificial intelligence, some care and assistance activities currently carried out by nursing staff will probably be performed by intelligent robots: but in care, humans can never be fully replaced by machines. Not only can humans appear irreplaceable for some activities (I am thinking, for example, of those situations where we feel the need to have at our side one who understands us and can empathise with our feelings), but it would still be up to us to programme even the most intelligent robots. In the light of these sci-fi scenarios that are now profiled on the horizon, it also becomes important for health operators to receive the most appropriate training to be ready to face the challenges the future holds. The question is not to fix whether new technology is morally acceptable, but for which tasks we can use intelligent systems without creating a risk for our humanity and at the same time promoting the good of the patient. Further, robots could be a significant resource in both everyday practice and training. Here is no incompatibility of principle between robot technology and care activity: on the contrary, with the help of robots we could cultivate that disposition to care and that empathy that will probably always remain our prerogative.

References

1. Borenstein, J., Pearson, Y.: Robot caregivers: harbingers of expanded freedom for all? Ethics Inf. Technol. **12**, 277–288 (2010)
2. Robin, B., et al.: Robotic assistants in therapy and education of children with autism: can a small humanoid robot help encourage social interaction skills? Univ. Access Inf. Soc. **4**, 105–120 (2005)
3. Vallor, S.: Carebots and caregivers: sustaining the ethical ideal of care in the 21st century. Philos. Technol. **24**(3), 251–268 (2011)
4. Coeckelbergh, M.: Health care, capabilities, and ai assistive technologies. Ethical Theory Moral Pract. **13**, 181–190 (2010)
5. Bouchard, S., et al.: Effectiveness of virtual reality exposure in the treatment of arachnophobia using 3D games. Technol. Health Care **14**(1), 19–27 (2006)
6. Garcia-Palacios, A., et al.: virtual reality in the treatment of spider phobia: a controlled study. Behav. Res. Ther. **40**, 983–993 (2002)
7. Huang, Z., et al.: Impact of using a robot patient for nursing skill training in patient transfer. IEEE Trans. Learn. Technol. **10**(2), 355–366 (2017)
8. Maidhof, W., Mazzola, N., Lacroix, N.: Student perceptions of using a human patient simulator for basic cardiac assessment. Currents Pharm. Teach. Learn. **4**(1), 29–33 (2012)
9. Moodley, T., Gopalan, D.: Airway skills training using a human patient simulator. Southern African J. Anaesth. Analg. **20**(3), 147–151 (2014)
10. Yinka, B.: Pediatric robot patient offers new level of realism for doctors in training. Medical Press, 10 September 2018. https://medicalxpress.com/news/2018-09-pediatric-robot-patient-realism-doctors.html
11. Tan, G.M., Ti, L.K., Suresh, S., Ho, B.S., Lee, T.L.: Teaching first-year medical students physiology: does the human patient simulator allow for more effective teaching? Singapore Med. J. **43**(5), 238–242 (2002)
12. Alinier, G., Hunt, B., Gordon, R., Harwood, C.: Effectiveness of intermediate-fidelity simulation training technology in undergraduate nursing education. J. Adv. Nurs. **54**(3), 359–369 (2006)
13. Crofts, J.F., Bartlett, C., Ellis, D., Hunt, L.P., Fox, R., Draycott, T.J.: Training for shoulder dystocia: a trial of simulation using low-fidelity and high-fidelity mannequins. Obstet. Gynecol. **108**(6), 1477–1485 (2006)
14. Cioffi, J., Purcal, N., Arundell, F.: A pilot study to investigate the effect of a simulation strategy on the clinical decision making of midwifery students. J. Nurs. Edu. **44**(3), 131–134 (2005)
15. La Cerra, C., et al.: Effects of high-fidelity simulation based on life-threatening clinical condition Scenarios on learning outcomes of undergraduate and postgraduate nursing students: a systematic review and meta-analysis. BMJ Open **9**(2) (2019). https://doi.org/10.1136/bmjopen-2018-025306
16. Persico, L.: A review: using simulation-based education to substitute traditional clinical rotations. JOJ Nurs. Health Care **9**(3), 1–7 (2018). https://doi.org/10.19080/JOJNHC.2018.09.555762

17. Sharkey, N., Sharkey, A.: The rights and wrongs of robot care. In: Lin, P., Abney, K., Bekey, G. (eds.) Robot Ethics: The Ethical ans Social Implications of Robotics, pp. 267–282. MIT Press, Cambridge (2011)
18. Wang, S., et al.: The uncanny valley: existence and explanations. Rev. Gen. Psychol. **19**(4), 393–407 (2015)
19. MacDorman, K.F., Ishiguro, H.: The uncanny advantage of using androids in cognitive and social Science research. Interact. Stud.: Soc. Behav. Commun. Biol. Artif. Syst. **7**, 297–337 (2006)
20. Saygin, A.P., Chaminade, T., Ishiguro, H., Driver, J., Frith, C.: The thing that should not be: predictive coding and the uncanny valley in perceiving human and humanoid robot actions. Soc. Cogn. Affect. Neurosci. **7**, 413–422 (2012)
21. Kanda, T., Hirano, T., Eaton, D., Ishiguro, H.: Interactive robots as social partners and peer tutors for children: a field trial. Hum.-Comput. Interact. **19**(1–2), 61–84 (2004)
22. Mori, M.: The uncanny valley (1970). Energy **7**(4), 33–35 (2012)
23. Mc Dorman, K.: Androids as an experimental apparatus: why is there an uncanny valley and can we exploit it? In: CogSci2005 Workshop: Toward Social Mechanisms of Android Science, pp. 108–118. Stresa (2005)
24. Carpenter, J.: Deus Sex Machina: loving robot sex workers and the allure of an insincere kiss. In: Danaher, J., McArthur, N. (eds.) Robot Sex. Social and Ethical Implications, pp. 261–287. The MIT Press, Cambridge (2017)

Virtual Reality for Informal Caregivers of Heart Failure Patients: A Mixed Method Research Proposal

Angela Durante[1]([⊠]) [iD], Antonella Palombo[1],
and Adriano Acciarino[2] [iD]

[1] Biomedicine and Prevention Department, University of Rome "Tor Vergata",
via Montpellier, 1, 00133 Rome, Italy
angela.durante@uniroma2.it
[2] Department of Psychology, University of Rome "La Sapienza",
via dei Marsi, 78, 00185 Rome, Italy
adriano.acciarino@uniroma1.it

Abstract. Informal Caregivers (CG) for heart failure patients are very important for preserving health status, preventing symptoms exacerbation and improving self-care. However, being a caregiver can be very burdensome, emotionally stressful and, often, associated with social interaction problems. Despite this, literature lacks in interventions aimed to reduce burden and improve wellbeing. The aim of this project is to reduce burden, improve mutuality and increase resilience with the use of immersive virtual reality. A concurrent mixed method design will be used to test the intervention with two experiments on volunteer CG. Open-ended questions will be used, meanwhile, to explore the impact of the experience. Caregiver Burden Inventory (CBI), Mutuality Scale (MS) and Connor Davidson Resilience Scale (CDRISC-25) will be administered both pre and post experiments. On psychophysiological indexes, an analysis of variance will be performed (ANOVA). Data about burden, mutuality and resilience will be presented through data integration with the content analysis of the open-ended questions in a joint display.

1 Introduction

1.1 Background

In the last years, literature about heart failure (HF) has increased focusing the attention on caregivers (CG) problems and needs [1–4]. Being a CG of a patient with a chronic disease is a hard role that changes during the day due to the multiple needs to cover; and can be easily burdensome along the time [5, 6]. Resilience and mutuality emerged as protective factors for the burdens of caregiving, to improve self-care and reduce hospitalizations [7–9]. Furthermore, it is not rare for CG to experience anxiety, depression and social isolation [10, 11]. The interventions available for support HF CG are still lacking, due to the recent research attention in this population and the underestimated social and economic value of their role.

© Springer Nature Switzerland AG 2020
E. Popescu et al. (Eds.): MIS4TEL 2019, AISC 1008, pp. 26–31, 2020.
https://doi.org/10.1007/978-3-030-23884-1_4

Parsons and Mitchell [12] underline the possible therapeutic use of Virtual Reality (VR) to improve social skills in individuals, through the simulations of real-world situations in a safe and controlled Virtual Environment (VE) [13, 14]. After that, also Fusaro and colleagues demonstrated the importance of the point of view in respect to a painful/pleasure stimulus delivered on a virtual hand experienced in VR [15]. A stronger electrodermal activity (EDA, analyzed as skin conductance response, SCR) was elicited when the point of view was in 1st person perspective (1PP), but when the avatar's hand was seen in a 3rd person perspective (3PP), the same stimulus, was perceived as less intense. The crucial effect of 1PP in a VE is to make an individual feel like his/her own body corresponds to the avatar's one, and it is immerged into a specific environment. This is due to the "Proteus effect", for which embodiment leads to shifts in self-perception, both online and offline, based on the avatar's features [16, 17]. Osimo and colleagues carried out another research in this field, creating a realistic avatar representing Sigmund Freud that participants could embody in 1PP or see him in 3PP into the VE. In the 1PP condition, they had in front of them their own-body avatar, and in the 3PP one, they were embodying their selves. In the second condition, participants had to talk about a personal problem, while in the first one they had to offer themselves a counselling. It was found that just when the counsellor resembled Freud, participants improved their general mood [18].

Avatar manipulation have the power to change implicit social biases by changing the relationship between the self and the outgroups through induction of ownership over bodies with different features (e.g., gender, age, ethnicity) compared to the participant's own real body [19]. A circular correlation between the sense of presence and emotions has been demonstrated [19]. To get into a VE with an avatar, which performs facial expressions, works as well as non-immersive procedures for the study of emotion recognition by facial expressions, especially when the avatars show naturalistic dynamic expression, changes [20]. In these studies, the feeling of presence was greater when an individual was immersed into an "emotional" VE (for example, an 'anxious' park vs a 'relaxing' one), and, on the other hand, the subjective emotional state was influenced by the level of presence. It has been also demonstrated that the avatars' emotional connotations can change the preference of participants on them, inducing an empathic response [21]. This capability of VR in inducing emotional states in a well-controlled way results in the Virtual Reality Exposure Therapy (VRET), an increasingly common treatment for Post-Traumatic-Stress-Disorder (PTSD) and anxiety [22].

VR methods allow researchers to manipulate participants' socio-affective dimension in a way that is almost impossible in the real environment, inducing changes in ingroup/outgroup prejudice and stereotypes, eliciting emotions and moods, and making people interact and collaborate in new ways.

To our knowledge, there is no VR application on CG of HF patients. In general, there is a lack of interventions in literature to reduce CGs' burden, despite the importance of the role, those people play for patients' wellbeing.

1.2 Aim

The aim of this project is threefold: (i) to reduce CGs' burden, (ii) to improve mutuality, and (iii) to increase their resilience.

2 Method

A concurrent mixed method design will be performed. Purpose sampling among CG of heart failure patients will be used. They must be more than 18 years old and able to sign a written consent. People with neurological disorders (e.g., epilepsy) will be excluded due to the use of VR to avoid possible negative side effects of the technique.

At the time of enrolment, each CG will be asked to complete a sociodemographic schedule (e.g. gender, age, education level, period of caregiving) and three scales for determine baseline burden, mutuality with the patient and resilience levels.

Burden will be measured using Caregiver Burden Inventory (CBI), developed based by Novak and Guest and psychometrically tested for HF caregiver in 2016 by Greco and colleagues [23, 24]. It is composed by 24 items clustered into five factors: time dependence, developmental, physical, social and emotional burden. Mutuality, the dyadic factor that expresses the interconnection between patients and CG, which represents the positive quality of the relationship, will be measured using the 15-item Mutuality Scale (MS). Archbold and colleagues found that caregivers with lower levels of mutuality experienced higher levels of caregiver role strain [25, 26]. To measure resilience Connor-Davidson Resilience Scale (CD-RISC25) will be used. It is a 25-item self-rating scale. The scoring of the scale is based on summing the total of all items, each of which is scored from zero to four [27].

Experiment 1: An avatar that looks like the patient who the CG care for will be created. Three pictures of the patient will be necessary to develop the avatars: one frontal and two profile pictures (left and right) of his/her face (using FaceGen, it is possible to paste patient's face pictures on a standard avatar). The avatar will have a neutral (more positive than negative) expression and will blink sometimes (to look more realistic). The CG will be immersed in a virtual environment (e.g., a neutral room), with patient's avatar staring at him/her. Around five minutes will be given to the CG to familiarize with the avatar and freely explore the whole environment. After free exploration, will be asked him/her to stay still, and Electrodermal Activity (EDA) and ECG will be recorded for 2 min. After the registration session, the CG will be asked to tell the avatar what kind of emotions (positive and/or negative) he/she feels toward it and what he/she likes and dislikes most of the patient. Then, he/she also will be asked to describe how he/she lives the role and the experience as CG, focusing on the activities across the day and the emotions elicited by them. At last, he/she will be asked the following question: "Feel free to tell the avatar something that you could never tell to the real patient. There are no time limits, so tell us when you're done".

Answers will be recorded and transcribed.

After immersion, the CG will be asked a feedback about his/her perceived immersion in the VE, the realism of patient's avatar, if he/she was really perceiving him/herself talking with the patient, and if it was easier to talk with the avatar compared to the real patient. All of these elements will be measured on a 7-point Likert scale.

Results will be computed on Excel data sheet and analyzed using SPSS21.

Experiment 2: Another 2 min recording of EDA and ECG on CG, will be performed during immersion, but in this second session, the CG will embody the patient's avatar in front of a virtual mirror. The avatar will be shown as dyspneic and with ankle swelling, simulating symptoms exacerbation. It will be asked the CG to stay still, and the recording will start. The CG will be asked to perform an everyday-life activity (e.g. to sit and lace up her/his shoes). Once finished, the CG will be asked about the experience and his/her personal feelings towards patient's condition.

Answers will be recorded and transcribed.

After the intervention, the CG will be asked to fill again the three scales about burden (CBI), mutuality (MS) and resilience (CD-RISC25).

2.1 Analysis

Descriptive statistics as mean and standard deviation or median and interquartile range will be used for continuous variables and n (percentage) will be used for categorical variables to summarize caregivers' demographic characteristic.

About psychophysiological indexes, an Analysis of Variance (ANOVA) will be perform after a pre-processing of both skin and heart signals. The open-ended questions asked to the CG during the experimental procedure will be analyzed performing a content analysis with NVIVO [28]. Data regarding Caregiver Burden Inventory, Mutuality Scale and Connor-Davidson Resilience Scale will be analyzed using SPSS21 and presented as data integration in a Joint Display with the results emerged from the interviews analysis [29].

3 Expected Results

A reduction of self-reported daily stress levels for the CG is expected on the CBI score. Moreover, a decrease of psychophysiological (EDA and HR activity) indexes of general stress is also supposed. Otherwise, an increase in CG comprehension and acceptance of patient's situation is predicted, in terms of MS and CD-RISC25 score, with benefit for dyadic relation. This could be the first step for the development of an "exposure therapy" for CG burden using IVR techniques.

References

1. Graven, L.J., Azuero, A., Grant, J.S.: Psychosocial factors related to adverse outcomes in caregivers of heart failure patients: a structural equation modeling analysis. J. Card. Fail. **24**, S101 (2018)
2. Grant, J.S., Graven, L.J.: Problems experienced by informal caregivers of individuals with heart failure: an integrative review. Int. J. Nurs. Stud. **80**, 41–66 (2018)
3. Nicholas Dionne-Odom, J., Hooker, S.A., Bekelman, D., et al.: Family caregiving for persons with heart failure at the intersection of heart failure and palliative care: a state-of-the-science review. Heart Fail. Rev. **22**, 543–557 (2017)
4. Wingham, J., Frost, J., Britten, N.: Behind the smile: qualitative study of caregivers' anguish and management responses while caring for someone living with heart failure. BMJ Open **7**, e014126 (2017)

5. Durante, A., Paturzo, M., Mottola, A., Alvaro, R., Vaughan Dickson, V., Vellone, E.: Caregiver contribution to self-care in patients with heart failure: a qualitative descriptive study. J. Cardiovasc. Nurs. (2018) https://doi.org/10.1097/jcn.0000000000000560

6. Hunt, G., Greene, R., Suridjan, C., Henningsen, N., Starr, M., Campbell, S., Arthur, G., Sadler, K., Strömberg, A.: Project group with representatives from the involved organizations: patient organizations and other supportive organizations: clinical researcher (2017)

7. Hooker, S.A., Schmiege, S.J., Trivedi, R.B., Amoyal, N.R., Bekelman, D.B.: Mutuality and heart failure self-care in patients and their informal caregivers. Eur. J. Cardiovasc. Nurs. **17**, 102–113 (2018)

8. Zhao, X., Lee, K., Baney, B., Penrod, J., Schubart, J.R.: Resilience in the context of informal care giving: a scoping study. Med. Res. Arch. **4**, 1–22 (2016)

9. Lum, H.D., Lo, D., Hooker, S., Bekelman, D.B. (2014). https://doi.org/10.1016/j.hrtlng.2014.05.002

10. Petruzzo, A., Paturzo, M., Naletto, M., Cohen, M.Z., Alvaro, R., Vellone, E.: The lived experience of caregivers of persons with heart failure: a phenomenological study. Eur. J. Cardiovasc. Nurs. **16**, 638–645 (2017)

11. Petruzzo, A., Biagioli, V., Durante, A., Gialloreti, L.E., D'Agostino, F., Alvaro, R., Vellone, E.: Influence of preparedness on anxiety, depression, and quality of life in caregivers of heart failure patients: testing a model of path analysis. Patient Educ. Couns. 13–15 (2018)

12. Parsons, S., Mitchell, P.: The potential of virtual reality in social skills training for people with autistic spectrum disorders. J. Intellect. Disabil. Res. **46**, 430–443 (2002)

13. Parsons, S., Cobb, S.: State-of-the-art of virtual reality technologies for children on the autism spectrum. Eur. J. Spec. Needs Educ. **26**, 355–366 (2011)

14. Cheng, Y., Huang, C.-L., Yang, C.-S.: Using a 3D immersive virtual environment system to enhance social understanding and social skills for children with autism spectrum disorders. Focus Autism Other Dev. Disabl. **30**, 222–236 (2015)

15. Fusaro, M., Tieri, G., Aglioti, S.M.: Seeing pain and pleasure on self and others: behavioral and psychophysiological reactivity in immersive virtual reality. J. Neurophysiol. **116**, 2656–2662 (2016)

16. Fox, J., Bailenson, J.N., Tricase, L.: The embodiment of sexualized virtual selves: the Proteus effect and experiences of self-objectification via avatars. Comput. Hum. Behav. **29**, 930–938 (2013)

17. Foucher, M.: L'Arc de crise, approche Française des conflits. Bull. d'Association Geogr. Fr. **89**, 6–17 (2012)

18. Osimo, S.A., Pizarro, R., Spanlang, B., Slater, M.: Conversations between self and self as Sigmund Freud—a virtual body ownership paradigm for self counselling. Sci. Rep. **5**, 13899 (2015)

19. Maister, L., Slater, M., Sanchez-Vives, M.V., Tsakiris, M.: Changing bodies changes minds: owning another body affects social cognition. Trends Cogn. Sci. **19**, 6–12 (2015)

20. Faita, C., Vanni, F., Lorenzini, C., Carrozzino, M., Tanca, C., Bergamasco, M.: Perception of basic emotions from facial expressions of dynamic virtual avatars, pp. 409–419. Springer, Cham (2015)

21. Rodrigues, S.H., Mascarenhas, S.F., Dias, J., Paiva, A.: "I can feel it too!": Emergent empathic reactions between synthetic characters. In: 3rd International Conference on Affective Computing and Intelligent Interaction and Workshops (2009)

22. Parsons, T.D., Rizzo, A.A.: Affective outcomes of virtual reality exposure therapy for anxiety and specific phobias: a meta-analysis. J. Behav. Ther. Exp. Psychiatry **39**, 250–261 (2008)

23. Novak, M., Guest, C.: Application of a multidimensional caregiver burden inventory. Gerontologist **29**, 798–803 (1989)

24. Greco, A., Pancani, L., Sala, M., Annoni, A.M., Steca, P., Paturzo, M., D'Agostino, F., Alvaro, R., Vellone, E.: Psychometric characteristics of the caregiver burden inventory in caregivers of adults with heart failure. Eur. J. Cardiovasc. Nurs. **16**, 502–510 (2017)
25. Archbold, P.G., Stewart, B.J., Greenlick, M.R., Harvath, T.: Mutuality and preparedness as predictors of caregiver role strain. Res. Nurs. Health **13**, 375–384 (1990)
26. Schumacher, K.L., Stewart, B.J., Archbold, P.G.: Mutuality and preparedness moderate the effects of caregiving demand on cancer family caregiver outcomes. Nurs. Res. **56**, 425–433 (2007)
27. Connor, K.M., Davidson, J.R.T.: Development of a new resilience scale: The Connor-Davidson Resilience Scale (CD-RISC). Depress. Anxiety **18**, 76–82 (2003)
28. Mayring, P.: Qualitative content analysis. Forum Qual. Sozialforsch./Forum Qual. Soc. Res. **1** (2000)
29. Cresswell, J.W.: Research Design: Qualitative, Quantitative, and Mixed Methods Approaches (2014). Google Books. https://doi.org/10.1007/s13398-014-0173-7.2

Efficacy of High-Fidelity Simulation on Learning Outcomes: Immediate Results for a Postgraduate Intensive Care Nursing Course

Angelo Dante[✉], Carmen La Cerra, Vittorio Masotta,
Valeria Caponnetto, Elona Gaxhja, Cristina Petrucci,
and Loreto Lancia

Department of Health, Life and Environmental Sciences,
University of L'Aquila, Edificio Delta 6 - Via Giuseppe Petrini,
67100 L'Aquila, Italy
angelo.dante@univaq.it,
{carmen.lacerra,vittorio.masottal,valeria.caponnetto,
elona.gaxhja}@graduate.univaq.it, {cristina.petrucci,
loreto.lancia}@cc.univaq.it

Abstract. High-Fidelity Simulation (HFS) founded on critical care scenarios requires nursing students to perform clinical interventions to effectively face life-threatening conditions. Since little evidence is available on the effectiveness of HFS on knowledge, self-confidence, self-efficacy, performance, and satisfaction of postgraduate nursing students attending an intensive care course, a before-and-after study was conducted on 28 students to document any change of learning outcomes in a respiratory failure scenario. After the HFS session, a significant improvement for self-confidence was revealed. Beyond the statistical significance, the HFS demonstrated to be a valuable complement to traditional education for many students, determining an improvement of their knowledge and self-efficacy. High level in performance and satisfaction during and after HFS session were also documented, confirming HFS as a valid teaching method in the achievement of learning goals, especially when associated with the traditional work placement models, in nursing students attending a postgraduate academic course on intensive and emergency care.

Keywords: High fidelity simulation training · Nursing students ·
Postgraduate Intensive Care Nursing course · Learning outcomes

1 Introduction

Among the most innovative uses of High-Fidelity Simulation (HFS), technologically improved manikins assume a key role since they can faithfully reproduce several clinical parameters, allowing students to learn and practice real clinical activities in a controlled and safe learning environment [1, 2]. In the nursing field, HFS-based education is increasingly founded on clinical scenarios that, consistently with expected clinical outcomes, require participants to perform a wide range of clinical interventions

© Springer Nature Switzerland AG 2020
E. Popescu et al. (Eds.): MIS4TEL 2019, AISC 1008, pp. 32–39, 2020.
https://doi.org/10.1007/978-3-030-23884-1_5

to face life-threatening conditions effectively [3, 4]. The use of HFS has shown to improve several nursing students' learning outcomes, among which knowledge and performance are the most investigated, as well as satisfaction, self-confidence, and self-efficacy [5, 6]. This feature has been confirmed particularly by available evidence derived from studies on undergraduate nursing students, whereas minor attention has been paid to the effectiveness of HFS on learning outcomes in postgraduate courses [4], although HFS utilizing critical care-based scenarios is going to be noticeably increased over the last years. During their traditional work placements in hospital and out-of-hospital clinical settings, nursing students could have a few chances to manage life-threatening conditions directly, and when this happens, errors potentially detrimental for both nursing students [7, 8] and patients can be made [9, 10]. For this reason, HFS seems to represent a valuable adjunct to the traditional clinical training in advanced nursing education to obtain better learning outcomes, especially in critical care area where rapid and effective interventions are often required [4]. Thus, considering the little evidence available in this field, this study aimed to explore the effectiveness of HFS based on critical care scenarios on the improvement of learning outcomes in students attending a postgraduate Intensive Care Nursing (ICN) course.

2 Methods

2.1 Study Design, Setting, and Population

A before-and-after study design was adopted during 2018 at the University of L'Aquila (Italy) where a 'Simulation Lab for Research Applied to Nursing Education' has been set. All nursing students who were attending a postgraduate ICN course, and who had expressed their written and informed consent were enrolled. No specific exclusion criteria were utilized. The features of Simulation Lab and postgraduate ICN course are reported in Table 1.

2.2 Variables and Measurement Tools

Information about sociodemographic data (gender, age), Bachelor of Science in Nursing (BSN) Degree grades, postgraduate course exam grades, current clinical work environment, and Basic Life Support (BLS) certification, were collected.

Knowledge and performance were defined as the 'level of the theoretical bases about caring' [11] and 'how students use their clinical skills' [12, 13], respectively. Self-efficacy was defined as 'how students think, feel, motivate themselves, and perceive their clinical practice' [14], self-confidence as 'how students believe in their own abilities when performing nursing care' [15], whereas satisfaction as 'the feeling of pleasure that students experience in their learning environment' [16].

Knowledge was measured with a 10-item *ad hoc* structured questionnaire exploring: (1) definition and causes of the clinical problem, (2) complications, pathophysiology, signs, and symptoms, and (3) nursing care. The score ranges from a minimum of 0 (lowest level) to a maximum of 10 (highest level). Performance was assessed through an 11-item *ad hoc* checklist aimed to document critical actions that guarantee a

Table 1. Features of Simulation Lab and postgraduate Intensive Critical Care course

Simulation Lab for Research Applied to Nursing Education
Simulation Room. The Simulation Room is equipped as a realistic Intensive Care Unit (ICU) room where the patient can be simulated by the wireless-controlled simulator (Gaumard® 'HAL® 1000'). Many part-task trainers allow students to train on specific clinical skills, e.g. endotracheal suction and intubation, suture of the skin, arterial cannulation. Finally, advanced audio/video recording devices are available.
Control Room. The Control Room, connected to the Simulation Room by a one-way mirror, is equipped with software-based controllers of the simulator and an audio/video system. This latter allows to record and live-stream simulation sessions to classrooms. Furthermore, the recorded scenarios can whenever be reproduced. From the Control Room is also possible to give the voice to the patient.
Briefing and Debriefing Room. A dedicated room equipped with an audio/video system allows to perform briefing and debriefing sessions. The possibility to replay the audio/video recording of the scenario running allows students to critically reflect on their practice and share their strengths and weaknesses.
Human Resources. Simulation Lab is directed by a Professor of Nursing. The design of the simulation scenarios and sessions are handled by a group of Researchers. Based on students training needs, other health care professionals are involved in designing and running simulation scenarios.
Postgraduate Intensive Care Nursing course
Training objectives. The general aim of the course is to give participants advanced nursing skills to allow them to provide best-quality care during the management of life-threatening conditions in each kind of critical care setting.
Admission and matriculation. BSN Degree is required to apply. An entry exam testing knowledge in the nursing field is mandatory to select a maximum of 40 students to attend the course.
Length, credits, and hours. 1 year, 60 credits (1 credit = 25 hours), 1500 hours.
Lectures. 30 credits of lectures are scheduled. Students must attend at least 70% of the lectures. Seminars are organized in order to deepen specific topics (e.g. Continuous Renal Replacement Therapy, Maxi emergencies, ECG interpretation).
Clinical training. Students are expected to attend 22 credits of clinical training in Critical Care Area or on rescue vehicles (e.g., ICU, Emergency Room, Ambulances). Students also access to the Simulation Lab in which they can repeatedly experiment complex clinical techniques.
Exams. Students are expected to pass 4 exams (one for each theoretical course).
Course organization. 1) President (Professor of Nursing); 2) Lecturers (Full and Associate Professors and in-charge Professors from the Italian NHS); 3) Clinical tutors (expert nurses) mentoring students during their clinical training.
Final exam. The discussion of a thesis is required to achieve the Master.

significant change in the patient's clinical evolution. A point is attributed to each accomplished critical action. The score ranges from 0 (no critical actions accomplished) to 11 (all critical actions accomplished). A cross-cultural adaptation of the General Self-Efficacy Scale (GSE) [17] was used to measure self-efficacy. This 10-item Likert-tool shows response options ranging from 1 (not true at all) to 4 (totally true) with the final score ranging from 10 (lowest) to a 40 (highest). For self-confidence and satisfaction, a cross-cultural adaptation of the Student Satisfaction and Self-Confidence in Learning (SSSCL) scale was used [18]. This 13-item Likert-instrument consists of the subscale measuring satisfaction (5 items) and the subscale measuring self-confidence (8 items), both showing response options from 1 (strongly disagree) to 5 (strongly agree). The overall satisfaction score ranges from 5 (lowest) to 25 (highest) points, and the overall self-confidence score ranges from 8 (lowest) to 40 (highest) points.

2.3 Procedures, Bias Control, and Statistical Analysis

Since postgraduate students routinely perform HFS sessions during their course, only the written consent to fill the questionnaires and to use the data for research aims was asked them. According to Italian Law [19], data were processed anonymously. Measurements of knowledge, self-confidence, and self-efficacy were taken before (T_0) and after (T_1) the simulation session, whereas performance and satisfaction were measured during and after the simulation, respectively. Students participated to a 60-min HFS session based on a validated and pre-tested acute respiratory failure scenario designed on real clinical data and considering the INACSL [20] and SIAARTI [21] guidelines, especially as regards the critical actions needed to manage this life-threatening condition effectively. In the briefing, the facilitator introduced students to the simulation session and available technologies. During the scenario running, participants were expected to perform all key actions to reach the best clinical outcomes (clinical stabilization) and prevent any undesirable events (e.g. respiratory and cardio-circulatory arrest). The simulator has been set to provide algorithm-based feedback in response to the actions made by the students. Moreover, 'on-the-fly' inputs, i.e. real-time changes along the algorithm, were admitted to react to unexpected actions or serious omissions carried out by the students. The debriefing aimed to focus on strengths and weaknesses shown by the students through the replay of audio/video recordings.

Measurement bias was limited by a cross-cultural adaptation process for GSE [17] and SSSCL [18] scales. Content and face validity analyses for knowledge and performance scales were also carried out. Furthermore, administration of questionnaires and scales took place in a dedicated room guaranteeing the same conditions of measurements for all participants. The measurement of performance was undertaken simultaneously by two researchers during the running scenario phase. Any disagreement between raters was solved by replaying the audio/video recordings.

Given the exploratory nature of the study, neither sample size calculation nor power analysis was performed. Participants characteristics, as well as performance and satisfaction levels, were described through frequencies and percentages for categorical variables and means and standard deviations (SDs) for continuous ones. Knowledge, self-confidence, and self-efficacy levels were described as means and SDs, as well as 'absolute' and 'normalized' learning gains [22, 23] calculated and defined as follows:

$$absolute\ learning\ gain\ = \frac{\text{Average } T_1 \text{ score}\% - \text{Average } T_0 \text{ score}\%}{\text{Max score}\%}$$

i.e. the ratio between learning gain (Average T_1 score% − Average T_0 score%) and the maximum score achievable;

$$normalized\ absolute\ gain\ = \frac{\text{Average } T_1 \text{ score}\% - \text{Average } T_0 \text{ score}\%}{100\% - \text{Average } T_0 \text{ score}\%}$$

i.e. the ratio between the learning gain (Average T_1 score% $-$ Average T_0 score%) and the 'maximum possible' learning gain (100% $-$ Average T_0 score%).

In addition, individual single-students' normalized learning gains [24], i.e. what the student achieved in tests, given what was possible for them to achieve, were calculated for all students and averaged as:

$$\frac{\sum \text{normalized absolute gain of students who showed a learning gain}}{\text{No. students who showed a learning gain}}$$

The values of the normalized gains were considered 'high' if $\geq$ 70.0%, 'moderate' if < 70.0% and $\geq$ 30.0%, 'low' if < 30.0% [23]. To document any significant differences between before and after HFS exposure, a paired t-test was calculated. Significance level was set at p $\leq$ 0.05. Data were analyzed using the IBM SPSS software version 19.0 (IBM Corp., Armonk, NY, USA).

3 Results

Out of 34 eligible students, 28 (82.3%) gave their consent to participate and were enrolled. Most of the participants were female (75.0%), and the entire sample had a mean age of 25.9 ($\pm$5.3) years, median 24.0 (IQR 3; range 22–49).

The average grade achieved in the BSN Degree was 104.9 ($\pm$6.6), median 107.0 (IQR 10; range 88–111), whereas the mean of the postgraduate course exam grades, at the moment of data collection, was 28.1 ($\pm$1.0), median 28.0 (IQR 1; range 25–30). Most of the participants worked in clinical settings other than a critical area (62.5%); in addition, almost all had a BLS certification (96.4%).

Mean scores of knowledge improved from 87.0% (8.7/10 $\pm$ 1.0) to 90.0% (9.0/10 $\pm$ 1.1) with 3.0% and 19.4% of absolute and normalized learning gains, respectively (Table 2).

Table 2. Participants' learning gains (n = 28)

		Knowledge	Self-confidence	Self-efficacy
T_0 score	Mean (SD)	8.7 (1.0)	32.0 (3.1)	28.5 (2.8)
	%	87.0	80.0	71.3
T_1 score	Mean (SD)	9.0 (1.1)	34.0 (3.2)	28.5 (4.7)
	%	90.0	85.0	71.3
Paired t-test	p-value	0.183	0.001	0.955
Absolute gain	%	3.0	5.0	0.0
Normalized absolute gain	%	19.4	25.0	0.0
Students with learning gain	%	35.7	71.4	39.3
Normalized individual single-students' learning gain	Mean	81.7	40.2	28.8

Ten participants (35.7%) showed real learning gain. Of these, the average of individual single-students' normalized learning gains was high (81.7%).

As regards self-confidence, the average scores significantly grew from 80.0% (32.0/40 ± 3.1) to 85.0% (34.0/40 ± 3.2) with an absolute gain of 5.0% and a normalized value of 25.0%. Globally, 71.4% of students (n = 20) showed a real gain on such outcome, while the average individual single-students' normalized learning gain was moderate (40.2%). Finally, the average levels of self-efficacy detected before and after the HFS session were the same (28.5/40); thus, no absolute learning gains emerged. However, in 39.3% of the students (n = 11), an increase of self-efficacy was pointed out showing the lowest single-students' normalized gains (28.8%).

As regards the assessed performance, the mean level was 75.5% (8.3/11 ± 1.5), median 72.7% (8; IQR 2; range 5–11); concerning the satisfaction in learning experience, the mean level was 91.2% (22.8/25 ± 2.3), median 94.0% (23.5; IQR 3; range 16–25).

4 Discussion and Conclusion

The enrolled sample reflected the distribution in the gender of Italian nurses [25], and the age of participants was in line with the minimum age expected for students attending a postgraduate ICN course. As regard learning outcomes, in accordance with the international literature [26], the self-confidence showed a significant improvement after the HFS session. Conversely, no significant learning gain emerged for knowledge and self-efficacy. In this regard, focusing the attention only on the statistical significance of the pre-post absolute differences in learning outcomes can lead to a limiting interpretation of the utility of these results in the educational field. In fact, because data related to the effects of HFS on the single-student remain hidden, the real potential of the HFS as a teaching strategy based on the individual approach could remain misunderstood. Consequently, moving the attention to the real possibility of learning gain, i.e. how much gain each student can achieve in every considered learning outcome compared to their starting point, allows to overcome this interpretative limit. In this regard, more than one-third of the participants obtained a real gain also for knowledge and self-efficacy, with a high- and low-gain magnitude, respectively. Therefore, beyond the statistical significance, the HFS demonstrated to be a valuable complement to the traditional education for many of the students investigated, determining an improvement of their learning outcomes. Concerning self-efficacy, HFS determined an immediate paradoxical effect on its levels; probably students adopt a very critical thinking about themselves when they perceive the relevance of the effects of their clinical interventions on critically ill patients. Reasonably, the fact of being exposed to a situation that is very similar to reality makes students aware of their need to improve their skills to satisfy the needs of critical patients better. Furthermore, based on Kolb's experiential learning theory [27], students may demonstrate significantly improved both knowledge and self-efficacy levels since, during a subjective time lapse after the immediate concrete learning session (HFS), they experience the reflective observation, abstract conceptualization, and active experimentation stages. In this study, learning outcomes have been measured immediately after simulation session, not allowing to capture any significant improvement in the above-mentioned outcomes supposed through Kolb's theory. In this regard, follow-up studies are needed to comprehend the impact of HFS on the knowledge and self-efficacy trend over time.

In agreement with available evidence, also in this study, performance [28] and satisfaction [29, 30] were positively associated to HFS exposure, confirming the validity of this teaching method, also in students attending a postgraduate critical care course. Its explorative nature, the lack of a control group, and the small sample size were the main limitations of this study. Moreover, the outcomes measurement was carried out immediately before and after the HFS sessions; therefore, it was not possible to document the learning outcomes trend over time.

Globally, HFS impacts positively on self-confidence levels in nursing students attending a postgraduate Intensive Care Nursing course. However, an individual approach shows an improvement in all the learning outcomes considered. HFS confirms to be a valuable teaching method in the achievement of learning goals, especially when associated with the traditional educational programs. Future studies should consider providing single-students' level data to allow educators to design customized educational strategies.

References

1. Waxman, K.: The development of evidence-based clinical simulation scenarios: guidelines for nurse educators. J. Nurs. Educ. **49**(1), 29–35 (2010)
2. Petrucci, C., La Cerra, C., Caponnetto, V., Franconi, I., Gaxhja, E., Rubbi, I., Lancia L.: Literature-based analysis of the potentials and the limitations of using simulation in nursing education. In: Vittorini, P., Gennari, R., Di Mascio, T., Rodríguez, S., De la Prieta, F., Ramos, C., Azambuja Silveira, R. (eds.) Methodologies and Intelligent Systems for Technology Enhanced Learning. AISC, pp. 57–64. Springer, Cham (2017)
3. McGaghie, W.C., Issenberg, S.B., Cohen, E.R., Barsuk, J.H., Wayne, D.B.: Does simulation-based medical education with deliberate practice yield better results than traditional clinical education? A meta-analytic comparative review of the evidence. Acad. Med. **86**(6), 706–711 (2011)
4. La Cerra, C., Dante, A., Caponnetto, V., Franconi, I., Gaxhja, E., Petrucci, C., Lancia L.: High-fidelity patient simulation in critical care area: a methodological overview. In: Di Mascio, T., Vittorini, P., Gennari, R., De la Prieta, F., Rodríguez, S., Temperini, M., Azambuja Silveira, R., Popescu, E., Lancia, L. (eds.) Advances in Intelligent Systems and Computing, pp. 269–274. Springer, Cham (2019)
5. Kim, J., Park, J.H., Shin, S.: Effectiveness of simulation-based nursing education depending on fidelity: a meta-analysis. BMC Med. Educ. **23**(16), 152 (2016)
6. Leigh, G.T.: High-fidelity patient simulation and nursing students' self-efficacy: a review of the literature. Int. J. Nurs. Educ. Scholarsh. **5**(1), 1–17 (2008)
7. Petrucci, C., Alvaro, R., Cicolini, G., Cerone, M.P., Lancia, L.: Percutaneous and mucocutaneous exposures in nursing students: an Italian observational study. Int. J. Nurs. Educ. Scholarsh. **41**(4), 337–343 (2009)
8. Dante, A., Natolini, M., Graceffa, G., Zanini, A., Palese, A.: The effects of mandatory preclinical education on exposure to injuries as reported by Italian nursing students: a 15-year case–control, multicentre study. J. Clin. Nurs. **23**(5–6), 900–904 (2014)
9. Newton, J.M., McKenna, L.: The transitional journey through the graduate year: a focus group study. Int. J. Nurs. Stud. **44**(7), 1231–1237 (2007)

10. Wolf, Z.R., Hicks, R., Serembus, J.F.: Characteristics of medication errors made by students during the administration phase: a descriptive study. J. Prof. Nurs. **22**(1), 39–51 (2006)
11. Hunt, D.P.: The concept of knowledge and how to measure it. JIC **4**(1), 100–113 (2003)
12. Garside, J.R., Nhemachena, J.Z.: A concept analysis of competence and its transition in nursing. Nurse Educ. Today **33**(5), 541–545 (2013)
13. Robb, Y., Dietert, C.: Measurement of clinical performance of nurses: a literature review. Nurse Educ. Today **22**(4), 293–300 (2002)
14. Zulkosky, K.: Self-efficacy: a concept analysis. Nurs. Forum **44**(2), 93–102 (2009)
15. Perry, P.: Concept analysis: confidence/self-confidence. Nurs. Forum **46**(4), 218–230 (2011)
16. Dante, A., La Cerra, C., Caponnetto, V., Franconi, I., Gaxhja, E., Petrucci, C., Lancia L.: Efficacy of high-fidelity patient simulation in nursing education: research protocol of 'S4NP' randomized controlled trial. In: Di Mascio, T., Vittorini, P., Gennari, R., De la Prieta, F., Rodríguez, S., Temperini, M., Azambuja Silveira, R., Popescu, E., Lancia, L. (eds.) Advances in Intelligent Systems and Computing, pp. 261–268. Springer, Cham (2019)
17. Schwarzer, R.: Measurement of Perceived Self-efficacy: Psychometric Scales for Cross-Cultural Research. Free University of Berlin, Institute for Psychology, Berlin (1993)
18. Jeffries, P.R., Rizzolo, M.A.: Designing and Implementing Models for the Innovative Use of Simulation to Teach Nursing Care of Ill Adults and Children: A National, Multi-Site, Multi-Method Study (2006)
19. Pub. L. No. 196: Codice in materia di protezione dei dati personali (2003)
20. INACSL Committee: INACSL standards of best practice: Simulation^{SM} Simulation design. Clin. Simul. Nurs. **12**, S5–S12 (2016)
21. Menarini, M., Petrini, F., Bigi, E., Donato, P., Di Filippo, A.: Linee-guida per la gestione preospedaliera delle vie aeree. Società Italiana di Anestesia Rianimazione e Terapia Antalgica. http://www.siaarti.it/SiteAssets/Ricerca/Linee-guida-per-la-gestione-preospedali e ra-delle-vie-aeree/linee_guida_file_32.pdf. Accessed 02 Oct 2019
22. Colt, H.G., Davoudi, M., Murgu, S., Rohani, N.Z.: Measuring learning gain during a one-day introductory bronchoscopy course. Surg. Endosc. **25**(1), 207–216 (2011)
23. Hake, R.R.: Interactive-engagement versus traditional methods: a six-thousand-student survey of mechanics test data for introductory physics courses. AJP **66**(1), 64–74 (1998)
24. McGowan, M., Davis, G.: Individual gain and engagement with teaching goals. In: Proceedings of the 26th Annual Conference of the International Group for the Psychology of Mathematics Education, North American Chapter OISE, Toronto (2005)
25. D'Addio, L.: La sanità italiana è donna. L'infermiere 2 (2011)
26. Boling, B., Hardin-Pierce, M.: The effect of high-fidelity simulation on knowledge and confidence in critical care training: an integrative review. Nurse Educ. Pract. **16**(1), 287–293 (2016)
27. Kolb, D.: Experiential Learning: Experience as the Source of Learning and Development. Prentice Hall, Upper Saddle River (1984)
28. La Cerra, C., Dante, A., Caponnetto, V., Franconi, I., Gaxhja, E., Petrucci, C., Alfes, C.M., Lancia, L.: Effects of high-fidelity simulation based on life-threatening clinical condition scenarios on learning outcomes of undergraduate and postgraduate nursing students: a systematic review and meta-analysis. BMJ Open **22**, e025306 (2019)
29. Baptista, R.C., Paiva, L.A., Goncalves, R.F., Oliveira, L.M., Pereira, M.F., Martins, J.C.: Satisfaction and gains perceived by nursing students with medium and high-fidelity simulation: a randomized controlled trial. Nurse Educ. Today **46**, 127–132 (2016)
30. Kang, K.A., Kim, S., Kim, S.J., Oh, J., Lee, M.: Comparison of knowledge, confidence in skill performance (CSP) and satisfaction in problem-based learning (PBL) and simulation with PBL educational modalities in caring for children with bronchiolitis. Nurse Educ. Today **35**(2), 315–321 (2015)

How to Reduce Biological Risk Among Nursing Students: A Research Project

Cristina Petrucci$^{(\boxtimes)}$, Valeria Caponnetto, Carmen La Cerra,
Vittorio Masotta, Elona Gaxhja, Angelo Dante, and Loreto Lancia

Department of Health, Life and Environmental Sciences, University of L'Aquila,
Edificio Rita Levi Montalcini - Via Giuseppe Petrini, 67100 Coppito,
L'Aquila, Italy
{cristina.petrucci,loreto.lancia}@cc.univaq.it,
{valeria.caponnetto,carmen.lacerra,vittorio.masottal,
elona.gaxhja}@graduate.univaq.it,
angelo.dante@univaq.it

Abstract. The risk derived by the transmission of biological agents (especially blood-borne pathogens) responsible for infectious diseases in human beings is called biological risk (or hazard). Biological risk is the most diffused occupational hazard in the hospital, and it represents a threat to nursing students particularly. This project aims to provide evidence of the efficacy of simulation laboratory in preventing (reducing) accidental exposure to potentially infected biological material among nursing students. Nursing students of the first, second and third years attending an Italian Nursing Degree Course, after they give their informed consent, will be enrolled in a longitudinal observational study to investigate the experience of nursing students about accidental exposures to biological materials before and after attending simulation training sessions with part task trainers, according to the "mastery learning" methodology. The mastery learning will be facilitated by audio-video recording during the students' performance in the high fidelity simulation laboratory. The study is expected to provide scientific evidence of the effectiveness of simulation training carried out by "mastery learning strategy" to prevent (reduce) accidental exposure to potentially infected biological material in a population of undergraduate nursing students.

Keywords: Nursing students safety · Part task trainers · Biological risk · Simulation laboratory · Mastery learning

1 Introduction

The risk derived by the transmission of biological agents (especially blood-borne pathogens) responsible for infectious disease in human beings is called biological risk (or hazard) [1]. Biological risk is the most diffused occupational risk in the hospital, which is responsible for 40–50% of all the occupational accidents [2–4]. The accidental biological exposures can occur by cutting, biting, splashing or dripping [5], and they are mostly represented by percutaneous and mucocutaneous injuries, i.e. sustained

© Springer Nature Switzerland AG 2020
E. Popescu et al. (Eds.): MIS4TEL 2019, AISC 1008, pp. 40–46, 2020.
https://doi.org/10.1007/978-3-030-23884-1_6

non-sterile exposures either in the form of needle-sticks, sharp object-related injuries or splashes of body fluids to the eyes and mucus membranes [6].

As many research works have highlighted, health profession students, and nursing students particularly, are at high risk of biological exposure. In an Italian study the biological risk was found to represent 95.8% of the total nursing students' occupational risk in hospital [7].

The RiBiSI (RIschio BIologico Studenti Infermieri, that is in English Biological Risk in Nursing Students) study conducted in 2008 (an observational study which, by means of a questionnaire, retrospectively investigated nursing students' exposures to potentially infected biological materials), showed that the 10.29% of the nursing students included in the survey (total sample 2215) experienced at least one accidental biological exposure during every year of hospital clinical training [8]. The results of RiBiSI 2 (an unpublished research with a sample of 427 subjects) performed six years later using the same methodology of RiBiSI showed even worse results: 38.6% of Nursing Students had been exposed to biological material during the same reference period. According to this last survey, nursing students experienced accidental exposure to blood or other potentially infected biological materials in the following circumstances: generic care, intravenous puncture, medication, sharp disposal, subcutaneous and intramuscular punctures, needle recapping, transport of biological material, sharp passing, intravenous catheter removal, vial opening, trichotomy and other procedures, such as the urinary catheterization. Accidental exposure to blood or other potentially infectious material could represent a life-changing experience [9].

So, how to reduce the biological risk among nursing students? It is important to activate learning strategies that make these students skillful before moving to hospital clinical placement [10]. Simulation-based learning could be an effective strategy to achieve this objective. Simulation training is defined as a highly customized interactive medium or program that allows individuals to learn and practice real-world activities in an accurate, realistic, safe and secure environment (MeSH - Year introduced: 2016). The international literature about simulation is focalized on the patient's safety [11–13], on improving clinical competence, confidence, and decreasing stress [14, 15], on cost reduction [16], but it lacks evidence on health care professionals' safety.

To our knowledge, so far just one study focused on the prevention of biological risk by simulation programs, demonstrating the impact of several interventions, including simulation training and procedure logs, on reducing the incidence of needle-stick injuries (NSIs) among U.S. medical students [17].

Therefore, the aim of the present project that we named SaNuSSiP (Safety of Nursing Students by Simulation Program) is to provide evidence on the efficacy of simulation laboratory to prevent (reduce) accidental exposure to potentially infected biological material among nursing students (Fig. 1).

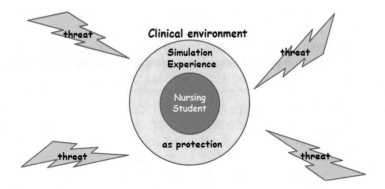

Fig. 1. Conceptual model guiding the project

2 Methods

2.1 Study Design and Population

All nursing students of the 1st, 2nd, and 3rd year attending an Italian nursing degree course (NDC) will be enrolled in a longitudinal observational study.

2.2 Setting

This study will be conducted involving students attending the NDC at the University of L'Aquila (central Italy). In Italy an NDC lasts 3 years and the degree is awarded upon completion of 180 credits (ECTS), which correspond to about 5400 h, 1800 of them in work placements. Clinical learning takes place in many different areas including medical, surgical, emergency, mental health and primary care services, where students are tutored by experienced registered nurses.

Since the hypothesis of this project is that simulation training can make nursing students safe from biological risk, the curricular work placement program will be integrated with simulation training sessions as recommended by the literature [18, 19].

An advanced clinical simulation laboratory equipped with part-task trainers, which are anatomical models of body parts in their normal state or representing disease, will be used. By part-task trainers in a free biological risk condition, students can train themselves and acquire skills such as:

- Intramuscular injection;
- Subcutaneous injection;
- Intravenous injection;
- Blood sampling;
- Enema;
- Urinary Catheterization.

Simulation training sections will be based on the "mastery learning" methodology [20]. The focus of "mastery learning" is on the standardization of the learning process rather than on the teaching process. Student must be able to demonstrate evidence of mastery before moving to the next section [21]. The mastery learning will be facilitated by debriefing sessions in which students can observe themselves while performing the procedures, having previously been audio-video recorded during their performance in the high fidelity simulation laboratory. Every skill will be considered achieved when the student will perform procedures without errors and will have reached a complete autonomy. Skills achieved by the students will be certified by the trained tutors.

2.3   Inclusion Criteria

Because of the ethical implications, all the students at 1st, 2nd, and 3rd year of the NDC will be included in the study considering simulation training sessions as a part of the regular curriculum.

2.4   Exclusion Criteria

Only students enrolled in Erasmus programs will be excluded.

2.5   Sample Size

Three hundred and twenty students, representing the whole target population, are expected to participate in the survey.

2.6   Development of the Research Project

At the beginning of the NDC 2nd and 3rd academic year (T0), a survey utilizing the RiBiSI tool will be carried out to investigate the experience of nursing students about accidental exposure to biological materials during the previous academic year, when they had attended just the traditional clinical work placement. These results will represent the comparison term (benchmark) to investigate the efficacy of the simulation training sessions on the students' accidental exposures to biological materials (Fig. 2 Research project timeline).

Then all the students (1st, 2nd, and 3rd year) will attend simulation training sessions according to the "mastery learning" methodology, and subsequently they will be moved in the real clinical settings for their regular placement.

At the end of the year course (T1), a survey utilising the same tool will be carried out to explore again the experience of nursing students about accidental exposure to biological materials that occurred during this period.

Results at T0 (1-year work placement survey NOT preceded by mastery learning simulation) will be compared with results at T1 (1-year work placement survey preceded by mastery learning simulation).

2.7   End-Points and Instruments

For the survey of this project, the same questionnaire used in previous surveys (RiBiSI) [8] and RiBiSI 2 will be used. The questionnaire consists of 27 multiple choice questions, focused on three areas:

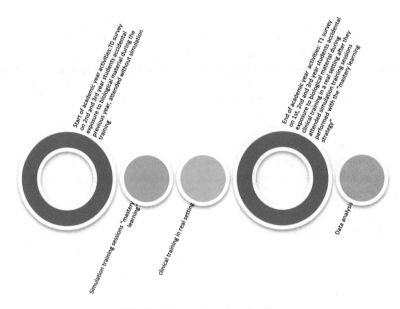

Fig. 2. Research project timeline

- demography (age, sex, year of study, out of course/in progress);
- events/outcomes (number, nature and circumstances of exposures);
- knowledge on biological risk (infections via blood, pre- and post-exposure prophylaxis, use of PPE).

The questionnaire will be filled anonymously by the students after their informed consent to participate to the surveys was given.

2.8 Data Management and Statistical Analysis

Demographic students characteristics will be collected, as well as the amount of accidental biological exposures normalized per hours of clinical training (exposition/training hours ratio) [8]. The chi-square test will be used to determine the differences between T0 and T1 results. Results will be expressed as Risk Relative (RR) with a 95% confidence interval (95% CI), and the statistical level of significance accepted as valid will be two-tailed p-values ≤ 0.05.

Data entry and processing will be done using the Statistical Package for Social Science (SPSS), version 19.0 software (IBM Corp., Armonk, NY, USA).

2.9 Strengths and Limitations

The strength but also the limitation of this project is represented by the reasons underlying the hypothesis of the study itself. In fact, although the demonstration of the effectiveness of an intervention would require an experimental design, as made in other studies [22], in this case such type of controlled trials would be unethical. So, from this consideration arises the need to apply the interventions on all students and to verify their effectiveness by comparing the results with those detected in the previous academic year, even if the subjects compared are different. However, the same methodology was used in similar studies [17].

2.10   Ethics

In agreement with the degree course board, ethical approval will be required by the Ethics Committee of the University of L'Aquila, according to National and European law [23, 24]. The students' participation in the survey will be voluntary after they are informed on the purpose, requirements, and duration of the study and after they give their informed consent in the terms approved by the Ethics Committee. Confidentiality will be maintained throughout the data collection and the data will be analyzed anonymously.

3   Expected Outcomes of the Study

The study is expected to provide scientific evidence on the effectiveness of simulation training carried out by "mastery learning strategy" to prevent (reduce) accidental exposure to potentially infected biological material in a population of Undergraduate Nursing Students.

If the research hypothesis is confirmed, an interesting contribution would have been made to the development of new safety work placement models.

References

1. Perry, J., Jagger, J.: Healthcare worker blood exposure risks: correcting some outdated statistics. Adv. Exp. Prev. **6**, 28–31 (2003)
2. Smith, D.R., Leggat, P.A.: Needlestick and sharps injuries among nursing students. J. Adv. Nurs. **51**(5), 449–455 (2005)
3. Yang, Y.H., Liou, S.H., Chen, C.J., Yang, C.Y., Wang, C.L., Chen, C.Y., Wu, T.N.: The effectiveness of a training program on reducing needlestick injuries/sharp object injuries among soon graduate vocational nursing school students in southern Taiwan. J. Occup. Health **49**(5), 424–429 (2007)
4. Stefanati, A., Boschetto, P., Previato, S., Kuhdari, P., De Paris, P., Nardini, M., Gabutti, G.: A survey on injuries among nurses and nursing students: a descriptive epidemiologic analysis between 2002 and 2012 at a University Hospital. Med. Lav. **106**(3), 216–229 (2015)
5. Samaranayake, L., Scully, C.: Needlestick and occupational exposure to infections: a compendium of current guidelines. Health Br. Dent. J. **215**(4), 163–166 (2013)
6. Bhattarai, S., Smriti, K.C., Pradhan, P.M., Lama, S., Rijal, S.: Hepatitis B vaccination status and needle-stick and sharps-related Injuries among medical school students in Nepal: a cross-sectional study. BMC Res. Notes **7**, 774 (2014)
7. Sacco, A., Stella, I.: Occupational injuries in nursing school students. Giornale Italiano di Medicina del Lavoro e Ergonomia **29**, 636 (2007)
8. Petrucci, C., Alvaro, R., Cicolini, G., Cerone, M.P., Lancia, L.: Percutaneous and mucocutaneous exposures in nursing students: an Italian observational study. J. Nurs. Scholarsh. **41**(4), 337–343 (2009)

9. Davenport, A., Cohn, S., Myers, F.: How to protect yourself after body fluid exposure. Nursing **39**(5), 22–28 (2009)
10. Ewertsson, M., Allvin, R., Holmström, I.K., Blomberg, K.: Walking the bridge: nursing students' learning in clinical skill laboratories. Nurse Educ. Pract. **15**(4), 277–283 (2015)
11. La Cerra, C., Dante, A., Caponnetto, V., Franconi, I., Gaxhja, E., Petrucci, C., Lancia, L.: High-fidelity patient simulation in critical care area: a methodological overview. In: Di Mascio, T., Vittorini, P., Gennari, R., De la Prieta, F., Rodríguez, S., Temperini, M., Azambuja Silveira, R., Popescu, E., Lancia, L. (eds.) Advances in Intelligent Systems and Computing, pp. 269–274. Springer, Cham (2019)
12. Petrucci, C., La Cerra, C., Caponnetto, V., Franconi, I., Gaxhja, E., Rubbi, I., Lancia, L.: Literature-based analysis of the potentials and the limitations of using simulation in nursing education. In: Vittorini, P., Gennari, R., Di Mascio, T., Rodríguez, S., De la Prieta, F., Ramos, C., Azambuja Silveira, R. (eds.) Methodologies and Intelligent Systems for Technology Enhanced Learning. AISC, pp. 57–64. Springer, Cham (2017)
13. Aggarwal, R., Mytton, O.T., Derbrew, M., Hananel, D., Heydenburg, M., Issenberg, B., MacAulay, C., Mancini, M.E., Morimoto, T., Soper, N., Ziv, A., Reznick, R.: Training and simulation for patient safety. Qual. Saf. Health Care **19**(Suppl 2), i34–i43 (2010)
14. Chen, S.H., Chen, S.C., Lee, S.C., Chang, Y.L., Yeh, K.Y.: Impact of interactive situated and simulated teaching program on novice nursing practitioners' clinical competence, confidence, and stress. Nurse Educ. Today **55**, 11–16 (2017)
15. Kiernan, L.C.: Evaluating competence and confidence using simulation technology. Nursing **48**(10), 45–52 (2018)
16. Zimmerman, D.M., House, P.: Medication safety: simulation education for new RNs promises an excellent return on investment. Nurs. Econ. **34**(1), 49–51 (2016)
17. Reid, M.J., Biller, N., Lyon, S.M., Reilly, J.P., Merlin, J., Dacso, M., Friedman, H.M.: Reducing risk and enhancing education: U.S. medical students on global health electives. Am. J. Infect. Control **42**(12), 1319–1321 (2014)
18. Burden, A., Pukenas, E.W.: Use of simulation in performance improvement. Anesthesiol. Clin. **36**(1), 63–74 (2018)
19. Issenberg, S.B., McGaghie, W.C., Petrusa, E.R., Lee Gordon, D., Scalese, R.J.: Features and uses of high-fidelity medical simulations that lead to effective learning: a BEME systematic review. Med. Teach. **27**(1), 10–28 (2005)
20. McGaghie, W.C.: Mastery learning: it is time for medical education to join the 21st century. Acad. Med. **90**(11), 1438–1441 (2015)
21. Dunn, W., Dong, Y., Zendejas, B., Ruparel, L., Farley, D.: Simulation, mastery learning and healthcare. Am. J. Med. Sci. **353**(2), 158–165 (2017)
22. Dante, A., La Cerra, C., Caponnetto, V., Franconi, I., Gaxhja, E., Petrucci, C., Lancia, L.: Efficacy of high-fidelity patient simulation in nursing education: Research protocol of 'S4NP' randomized controlled trial. In: Di Mascio, T., Vittorini, P., Gennari, R., De la Prieta, F., Rodríguez, S., Temperini, M., Azambuja Silveira, R., Popescu, E., Lancia, L. (eds.) Advances in Intelligent Systems and Computing, pp. 261–268. Springer, Cham (2019)
23. Decreto legislativo 30 giugno 2003, n. 196. Codice in materia di protezione dei dati personali. Gazzetta Ufficiale n. 174 del 29-7-2003 - Suppl. Ordinario n. 123 (2003)
24. Decreto legislativo 10 agosto 2018, n. 101. Disposizioni per l'adeguamento della normativa nazionale alle disposizioni del regolamento (UE) 2016/679 del Parlamento europeo e del Consiglio, del 27 aprile 2016, relativo alla protezione delle persone fisiche con riguardo al trattamento dei dati personali, nonché alla libera circolazione di tali dati e che abroga la direttiva 95/46/CE (regolamento generale sulla protezione dei dati). (18G00129) Gazzetta Ufficiale Serie Generale n. 205 del 04-09-2018 (2018)

Electronic Test of Competence Administration: Qualitative Evaluation of Students' Satisfaction on Telematic Platform a Cross Sectional Study

Albina Paterniani[1(✉)], Ilaria Farina[1], Giovanni Galeoto[1],
Claudia Quaranta[2], Francesca Sperati[3], and Julita Sansoni[1]

[1] Nursing Research Unit - Public Health and Infectious Disease,
Sapienza University of Rome, Piazzale Aldo Moro 5, 00185 Rome, Italy
albina.paterniani@uniroma1.it
[2] Sapienza University of Rome, Piazzale Aldo Moro 5, 00185 Rome, Italy
[3] IFO – Regina Elena National Cancer Institute, Rome, Italy

Abstract. Italian universities have introduced the Progress testing (PT) for a longitudinal assessment of student knowledge retention. This process provides educators a valid and reliable tool to assess learning while providing students measurement of competencies and self-evaluation of gaps in knowledge. The aim of the study is to evaluate the satisfaction of the students over three years in the nursing bachelor's degree program at the Sapienza University of Rome, on the compilation of the Electronic Test of Competence (TECO) through the use of an instrument evaluation. The sample was recruited from October 2018 to December 2018 at Sapienza University of Rome. The population we investigated consists of 250 students; results indicate TECO useful for monitoring preparation of the students of degree course in nursing, evaluating the degree courses in nursing and for the evaluation of the degree program. In conclusion we can affirm that the students are partly satisfied with the TECO telematic administration, and evidentiated some doubts on the first part of final graphics.

Keywords: Telematic platform · Nursing · Electronic competence test · TECO

1 Introduction

Italian universities have introduced the Progress testing (PT) for a longitudinal assessment of student knowledge retention. The tool used is an Electronic Test of Competence called TECO. This process provides educators a valid and reliable tool to assess learning while providing students the measurement of their competencies and self-evaluation of gaps in knowledge the process itself increases student satisfaction [1, 2]. Internationally, student satisfaction is considered pivotal for assessing positive connection between motivation and long-term learning [3].

Contents from any stage of nursing curriculum can appear in PT; therefore PTs should promote meaning-orientated learning and provide students with the grade of their retention during the studies [4]. Many Institutions all over the world have adopted

© Springer Nature Switzerland AG 2020
E. Popescu et al. (Eds.): MIS4TEL 2019, AISC 1008, pp. 47–54, 2020.
https://doi.org/10.1007/978-3-030-23884-1_7

progress testing, continuing to improve nursing and medical education [5]. The National Agency for the Evaluation of Universities and Research Institutes (ANVUR 2012) started a project on to evaluation university and student.

TECO, is important tool to monitor the quality of the education process, as indicated by the European Higher Education Area (EHEA) where the focus are students and competencies (European Standards and Guidelines for Quality Assurance, 2015).

The Ministerial Decree 987/2016 underlines the importance of TECO as indicator (DPR 76/2010 art.3) for university didactic process evaluation (Self-Assessment, Periodic Evaluation, Accreditation - AVA). After revision in 2016, it was redesigned introducing transversal competencies (TECO-T) and disciplinary competencies (TECO-D). Transversal competencies refer to student's ability in literacy and numeracy. The first two can compare the ability of students in different courses of study, while the disciplinary competencies focus on nursing discipline.

Participation to TECO is voluntary. The results are communicated individually to students and in a summative form to educators. Results do not influence in progress or final evaluation. The objectives of the disciplinary TECO are: to reach a shared definition of the disciplinary contents by declining them with the 5 Dublin Descriptors (knowledge and understanding, application of knowledge and understanding, elaboration of judgments, communication skills, learning abilities), self-evaluation of individual CdS, management centralized and certified (CINECA) for the administration and collection of data [5].

Galeoto et al. 2018, assessed the satisfaction of the TECO in paper format, resulting in simple, clear, and understandable, but difficult in answering questions due to a lack of practicality in the paper test [6].

The aim of the present study is to evaluate the satisfaction of the students over three years in the nursing bachelor's degree program at the "Sapienza" University of Rome, on the compilation of electronic TECO test.

2   Materials and Methods

2.1   Population

Bachelor degree in Nursing students. The sample was recruited from October 2018 to December 2018 at Sapienza University of Rome, on voluntary basis of participation. In order for the students to be included in the study, they had to meet the following inclusion criteria: (1) Enrolled in three-years degree program in nursing or graduating students of the November- December session of nursing courses offered in the Italian language. Exclusion criteria: (1) Students not enrolled in a three-years degree program in nursing; (2) Students enrolled in nursing courses offered in English.

2.2   Instrument

A Satisfaction ad hoc questionnaire [6]; was composed by nine questions: six multi-choice and three likert scale was used (Table 2). The questionnaire, under scientific validation, was distributed to the university platform on voluntarily and anonymous base that included demographic information [7].

Electronic TECO

The TECO project aims to construct indicators that reflect the skills developed from the first through the third year of the university degree. The generic/transversal competencies defined by ANVUR are Literacy and Numeracy. The research hypothesis is that these competencies draw on general formative education and are taught throughout the term of higher education, assuring comparable information between institutions and program studies. The disciplinary competencies, by contrast, are closely linked to the specific educational contents of the students' program and consequently can be compared only with those from similar programs.

TECO-Cineca Platform

A CINECA institutional platform was created for the administration of the TECO on skills. The platform includes three persons involved in the process of creation/administration of the test:

- University contact: the person who prepares the sessions for the administration of the test
- Classroom tutor: the person who opens and closes the administration of the test
- Student: who compiles and sends the test data

Prerequisites

Students informative guidelines by Universitaly must be followed by students before to seat the test (identity document, tax code and Universitaly credentials).

Student Activity: Operating Instructions

An essential requirement for the test is the registration on the Universitaly website. If the student is not registered he can do it directly by clicking on the blue "Subscribe to the test" box on the home page (Fig. 1). The student must access the area: https://verificheonline.cineca.it/teco/login_studente.html.

Fig. 1. Page of student

Through the "Subscribe" button, students who are not registered on the Universitaly portal will have to fill in the regular form of subscription to Universitaly (first screen) and then the registration at the test (second screen) (Fig. 2).

Fig. 2. Registration for the test

Access to the Test

Once enrolled at Universitaly, the student (in the classroom) will have to access the private area: https://verificheonline.cineca.it/teco/login_studente.html Click on Subscribe to the Test (blue box) and access the test by entering your Fiscal Code (CF) and the password defined by Class Tutor (Fig. 3).

Fig. 3. Start test

Times of Administration

The session of the TECO-T is 75 min in total: 35 min for the Literacy module and 40 for the Numeracy module. After the TECO-T, the students answer the TECO-D questions in a session lasting 90 min. Enrollment by students at the Universitaly site takes approximately 10–15 min.

3   Statistical Analysis

The categorical variables have been reported through absolute and relative frequencies, the continuous variables by means of averages and standard deviations (DS). We evaluated the association of questions related to TECO satisfaction with gender, median age (≤ 20 vs >20 years) and the academic year (1 vs 2–3) using the Chi-2 test. We considered statistically significant all the differences with p-value <0.05. Data were analyzed using SPSS statistical software version 21 (SPSS inc., Chicago IL, USA).

4   Results

The sample consists of 250 students over three years in the nursing bachelor's degree program at the Sapienza University of Rome, the socio-demographic characteristics are described in Table 1.

Table 1. Characteristics of sample (250 students)

Male gender	77/250 (30.8%)
Age in years	21.39 ± 4.38
1st year students	203/250 (81.2%)
2nd year students	32/250 (12.8%)
3rd year students	15/250 (6.0%)

The Electronic TECO is difficult to answer in 75.6% of the sample; 53,2% claim for clarity and 53,6% consider partially understandable. 38.8% is not satisfied with graphic presentation; 43,2% is satisfied with style and the 76,4% consider the time given to full fill the questionnaire good. A moderately effective value was found for 36.0% in relation to the treatment of the students, the evaluation of the CL for 39.6% and finally in relation to the evaluation on the individual preparation 42.8% (Table 2).

Table 2. Descriptive analysis of questionnaire

Sample=250				
Do you think it is easy to answer the questions of the paper/electronic TECO?				
No	*In part*	*Yes*		
43 (17.2%)	189 (75.6%)	18 (7.2%)		
Are the contents of the paper/electronic TECO clear?				
14 (5.6%)	133 (53.2%)	103 (41.2%)		
Did you understand the contents of the paper/electronic TECO?				
9 (3.6%)	134 (53.6%)	107 (42.8%)		
Did you like the way the paper/electronic TECO was graphically designed?				
97 (38.8%)	82 (32.8%)	71 (28.4%)		
Are you satisfied with the style of the paper/electronic TECO?				
43 (17.2%)	99 (39.6%)	108 (43.2%)		
In your opinion, is the time sufficient to compile the paper/electronic TECO?				
15 (6.0%)	44 (17.6%)	191 (76.4%)		
In your opinion, how useful is the TECO for monitoring the preparation of students in the nursing-degree program?				
Not	*Slightly*	*Moderately*	*Very*	*Extremely*
34 (13.6%)	57 (22.8%)	90 (36.0%)	60 (24.0%)	9 (3.6%)
In your opinion, how useful is the TECO for evaluating the degree courses in the nursing-degree program?				
35 (14.0%)	63 (25.2%)	99 (39.6%)	42 (16.8%)	11 (4.4%)
In your opinion, how useful is the TECO for evaluation of your degree-course preparation?				
43 (17.2%)	46 (18.4%)	107 (42.8%)	39 (15.6%)	15 (6.0%)

Items statistically significant through stratification by gender, age and year of course are reported in Table 3.

Question "**Are you satisfied with the style of the TECO?**" Showed that students aged ≤ 20 years are partially satisfied for 42.9% (N 66) while students >20 years are satisfied for 54.2% (N 52) this data is statistically significant $p < 0.05$; in relation to the academic year the 1 year is not always satisfied for 41.9% (N 85) and the 2–3 year students are satisfied for 59.6% (N 28) this figure is statistically significant p value < 0.05.

Question "**In your opinion, how useful is the TECO for monitoring the preparation of students in the nursing-degree program?**" showed that 67.5% (N 104) of the sample ≤ 20 year and 44.8% (N 43)> 20 year states that the TECO is moderately effective, this data is statistically significant p-value 0.05; in relation to the academic year the students of the 1 year say for 63.1% (N 128) to be moderately effective and the students of the 2–3 year for the 44.7% (N 21) extremely effective this datum is statistically significant p-value 0.05;

Question "**In your opinion, how useful is the TECO for evaluating the degree courses in the nursing-degree program?**" in relation to age showed that students ≤ 20 year for 70.8% (N 109) and >20 year for 55.2% (N 53) evaluate the moderately effective TECO this figure is statistically significant p-value 0.05.

Table 3. Stratification of sample by gender, age and year of course

	Gender		p	Median Age		p	Academic year		p
	Male	Female		≤20 year	>20 year		1 N=203	2-3 N=47	
	N=77	N=173		N=154	N=96				
D5: Are you satisfied with the style of the TECO?									
No	14 (18.2%)	29 (16.8%)		32 (20.8%)	11 (11.5%)		38 (18.7%)	5 (10.6%)	
In part	28 (36.4%)	71 (41.0%)	0.78	66 (42.9%)	33 (34.4%)	0.02*	85 (41.9%)	14 (29.8%)	0.04*
Yes	35 (45.5%)	73 (42.2%)		56 (36.4%)	52 (54.2%)		80 (39.4%)	28 (59.6%)	
D7: In your opinion, how useful is the TECO for monitoring the preparation of students in the nursing-degree program?									
Not effective	14 (18.2%)	20 (11.6%)		20 (13.0%)	14 (14.6%)		27 (13.3%)	7 (14.9%)	
Slighty/ Moderately	41 (53.2%)	106 (61.3%)	0.31	104 (67.5%)	43 (44.8%)	0.01*	128 (63.1%)	19 (40.4%)	0.01*
Very/ Extremely	22 (28.6%)	47 (27.2%)		30 (19.5%)	39 (40.6%)		48 (23.6%)	21 (44.7%)	
D8: In your opinion, how useful is the TECO for evaluating the degree courses in the nursing-degree program?									
Not effective	12 (15.6%)	23 (13.3%)		20 (13.0%)	15 (15.6%)		26 (12.8%)	9 (19.1%)	
Slighty/ Moderately	48 (62.3%)	114 (65.9%)	0.84	109 (70.8%)	53 (55.2%)	0.03*	138 (68.0%)	24 (51.1%)	0.09
Very/ Extremely	17 (22.1%)	36 (20.8%)		25 (16.2%)	28 (29.2%)		39 (19.2%)	14 (29.8%)	

5   Discussion

The Progress Test is internationally widely used to assess university skills and knowledge [3]. In Italy, the Test of Competencies (TECO) is an important tool for monitoring the quality of the educational process and it is focused on student and related skills.

In general satisfaction of students from nursing degree courses after TECO administration by telematic platform was considered moderate. Three specific item about TECO style and usefulness show a partial satisfaction requiring future discussion and revisions. These data are important to evaluate and review nursing curriculum at difference level of the University educational process.

TECO is useful for monitoring preparation of students and to evaluate University degree program. The sample (250) was stratified by age; some questions were unclear in the cluster under 20 years old. This data need a further evaluation to understand if it is due to a lack of knowledge among first year students of the nursing bachelor course.

In other cases the font used is too small and for this reasons difficult to associate the image with the correct answer (Table 3). The students attending the first year, found more difficulties then others: a discussion is necessary to discover the related item because the TECO questions are the same for three years and the answer is depending by the knowledge acquired.

The sample is satisfied with the TECO style. Students consider the on going evaluative project useful particularly because the replication of the same test in the different administration gives them the measure of their knowledge and the awareness of their individual path. Student consider the time given to fill the test good.

In conclusion we can confirm the usefulness of telematic administration of TECO as a tool to evaluate students knowledge progress, to discuss nursing program offered by our University and to improve them.

References

1. Blake, J.M., Norman, G.R., Keane, D.R., Mueller, C.B., Cunnington, J., Didyk, N.: Introducing progress testing in McMaster University's problem based medical curriculum: psychometric properties and effect on learning. Acad. Med. (1996)
2. Albanese, M., Case, S.: Progress testing: critical analysis and suggested practices. Adv. Health Sci. Educ. (2015)
3. Sum, V., McCaskey, S.J., Kyeyune, C.: A survey of satisfaction levels of graduate students enrolled in a nationally ranked top-10 program at a mid-western university. Res. High. Educ. J. (2010)
4. Chen, Y., Henning, M., Yielder, J., Jones, R., Wearn, A., Weller, J.: Progress testing in the medical curriculum: students' approaches to learning and perceived stress. BMC Med. Educ. (2015)
5. Neeley, M., Ulman, A., Sydelko, S., Borges, J.: The value of progress testing in undergraduate medical education: a systematic review of the literature. Med. Sci. Educ. **26**, 617–622 (2016)
6. Galeoto, G., Rumiati, R., Sabella, M., Sansoni, J.: The use of a dedicated platform to evaluate health-professions university courses. In: Methodologies and Intelligent Systems for Technology Enhanced (2019)
7. http://www.anvur.it/wp-content/uploads/2018/05/Risultati-progetto-TECO. Accessed 15 Nov 2018

The Validity of Rasterstereography as a Technological Tool for the Objectification of Postural Assessment in the Clinical and Educational Fields: Pilot Study

Anna Berardi[1], Giovanni Galeoto[2]([⊠]), Marco Tofani[3],
Massimiliano Mangone[4], Serena Ratti[1], Arianna Danti[1],
Julita Sansoni[2], and Maria Auxiliadora Marquez[5]

[1] Sapienza Università di Roma, Rome, Italy
[2] Department of Public Health and Infectious Diseases,
Sapienza University of Rome, Rome, Italy
giovanni.galeoto@uniroma1.it
[3] Neurorehabilitation Unit, Bambino Gesù Children's Hospital, Rome, Italy
[4] Department of Anatomy Histology, Forensic Medicine and Orthopedics,
Sapienza University of Rome, Rome, Italy
[5] CPO – Paraplegic Center of Ostia Rome, Rome, Italy

Abstract. The present study aimed to validate Formetric 4D with wheelchair users in a sitting position, comparing the results with the data obtained from a postural assessment. Nine individuals with spinal cord injury were evaluated through postural assessment. Each individual was subjected to rasterstereography in a sitting position using the Formetric 4D. Test-retest reliability was evaluated, after four hours: at first, the detection was performed on the same participant three times by three different rater. After four hours, the detection was repeated by the first operator. This study highlighted the ability of Formetric 4D to provide consistent results at different times and with different evaluators, reporting a Cronbach's alpha of 0.74 and excellent intra- and inter-operator stability with an ICC of 0.91 to 0.96. Compared to the postural evaluation, the degree of accuracy of the measurements acquired with Formetric 4D, analyzed through Spearman's rho, showed statistically significant positive correlations with anthropometric measurements. The present study provides information enabling the use of the Formetric 4D tool in clinical, research, and educational settings this will be a very useful tool that allows students to have a three-dimensional representation of the anatomical components involved in the sitting position, helping them to learn and gain an in-depth understanding of how to perform an objective postural assessment examination.

Keywords: Formetric 4D · Posture · Assessment · Spinal cord injury

1 Introduction

Seating and postural management describes the way in which our bodies are positioned and managed. People with spinal cord injury, for whom a wheelchair is the means to move and perform their daily activities, often have postural changes due to the absence

The original version of this chapter was revised: The author names have been updated. The correction to this chapter is available at https://doi.org/10.1007/978-3-030-23884-1_20

E. Popescu et al. (Eds.): MIS4TEL 2019, AISC 1008, pp. 55–62, 2020.
https://doi.org/10.1007/978-3-030-23884-1_8

of mobility and muscle recruitment [1]. An obligatory static posture requires periodic checks to assess and monitor the patient's condition over time to prevent pressure injuries and the aggravation of deformities and avoid tendon and joint retractions [2]. Manual postural assessment is the method currently used by occupational therapists to evaluate these conditions, following the International Organization for Standardization's (ISO) standard number 16840:2013-2018 [3, 4] Postural assessment, however, is a subjective instrument and difficult to understand for students of occupational therapy, because it requires the therapist to record movements or joint deformity through palpation and mobilization. As a result of inadequate professional training, there is a scarcity of rehabilitation professionals experienced in or specially trained to provide seating and mobility recommendations [5]. Although the need to train additional skilled practitioners is clear, the most effective means of training and the tools to evaluate the effectiveness of training programs are not yet clear. Therefore, it is necessary to identify an instrument that provides scientifically valid data for an overall assessment of the patient's posture that can be repeated over time without causing harmful effects to the subject, giving students an instrument that enables them to see in a direct and three-dimensional way the components of the sitting position so that they can compare those components with the results of the postural evaluation. In the literature, several studies have demonstrated the reliability and validity of the rasterstereography performed by the Formetric 4D tool with the patient in a standing station, validating it as a tool for follow-up and screening [6, 7] The Formetric 4D tool facilitates clinical practice by analyzing the spinal column. It is completely radiation-free and is a non-invasive method that has been shown to be highly correlated with radiography and is extremely usable and reproducible.

It is essential to have an instrument that provides scientifically valid data for an overall assessment of the patient's posture that can be easily used by nursing and rehabilitation sciences students. The present study aims to evaluate for the first time Formetric 4D [8, 9] reliability (internal consistency, intra-rater and inter-rater reliability) in a sitting position, and evaluate its validity comparing the results with the anthropometric measurements (through the Spearman's rho correlation coefficient) detected during the postural evaluation.

2   Material and Methods

2.1   Population and Procedures

To conduct this study, individuals were recruited from the spinal unit of the Paraplegic Center in Ostia, Italy between June and September 2018. According to the established inclusion criteria, they needed to be over eighteen years old, have a spinal injury at the lumbar spine level, have no pressure injuries during the study, and use a wheelchair more than 50% of the time each day. Individuals who had pressure lesions, pregnant women, and patients in a state of entrapment were excluded from the study. As a preliminary step (t0), an interview was carried out with each participant during which every person was specifically informed about the procedures and aims of the study; personal data necessary for the research were collected after the acquisition of informed consent. In the first phase of the study (t1), the participants were evaluated by an occupational therapist through postural assessment. In the second phase (t2), each individual was evaluated with rasterstereography in a sitting position using the Formetric 4D by three different raters at the Policlinico Umberto I, Rome, Italy. With the assistance of two operators, the user transferred from the wheelchair to a stool that was adjustable in height. The same positioning was performed for all subjects. The third phase (T3) consisted of the test-retest, carried out after about four hours: at first, the detection was performed on the same participant three times by three different rater. After four hours, the detection was repeated by the first operator. This made it possible to establish intra-operator reliability between the tests and re-tests and inter-operator reliability measured on the same participant at the same time by the three different operators.

2.2   Instruments

The postural assessment procedures is developed from the ISO 16840:2013-2018 standards, which permits the determination and recording of a person's posture while seated in a wheelchair; the standard terms and definitions for use in describing both the posture and the anthropometrics of a person seated in a wheelchair. (3, 4) The purpose of this assessment procedure is to provide step by step directions to perform assessment tasks, and gather and analyze information when recommending seating and/or mobility equipment. The postural assessment is divided into three parts: observation of the person in current seating system, assessment in the supine position, and recording of anthropometric measurements. Prior to assessing current sitting posture, optimal positioning in the chair is obtained. The position of the participant in a wheelchair is observed to acquire the position of the head, limbs, trunk, and pelvis under normal conditions, that is, how the subject positions him- or herself in his or her daily routine. Subsequently, the participant is asked to lie down on the firm surface in a supine position and to align the body, head and pelvis (to neutral). The mobility of the pelvis is observed; in particular, for this study, the movements of pelvic tilt/lumbar lordosis, pelvic rotation, and pelvic obliquity were investigated. In order to recorder the anthropometric measurements during the sitting evaluation if the person's pelvis tends to assume a posterior or anterior pelvic tilt/pelvic obliquity/pelvic rotation, and he/she is unable to independently move his/her pelvis to neutral, the flexibility is assessed passively. To perform the rasterstereography, through the Formetric 4D tool (Fig. 1), the participant positioned him- or herself for a few seconds two meters from the detection device that uses halogen light projected on the participant's back in the form of a special horizontal line grid; the light is detected by a digital camera.

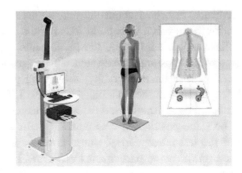

Fig. 1. Formetric 4D tool for rastesterography

Thanks to this optical scan, the system automatically detects anatomical landmarks, represented by C7 or prominent cervical vertebra, sacrum and lumbar dimples, and the median or symmetry line. The software analyzes the data thus obtained and reconstructs the shape of the back, the column, and the position of the basin in three dimensions: it creates 12 images in 6 s, calculating and representing the average value. Moreover, thanks to the three-dimensional reconstruction, the scan is carried out only on the posterior body surface; therefore, the subject does not have to reposition for the analysis on the other planes, thus minimizing the effect of postural variations.

2.3   Statistical Analysis

The statistical analysis of the study results was performed using SPSS software. The internal consistency of the Formetric 4D was measured with Cronbach's alpha. The intraclass correlation coefficient (ICC) was calculated to evaluate the intra-rater (the same operator t2–t3) and inter-rater reliability (three different operators t2) of the instrument. The correlation coefficient Spearman's rho was used to measure the correlation between the data obtained with the rasterstereography and the anthropometric measurements detected during the postural evaluation. Statistical significance was established for a p less than 0.05.

3   Results

Ten participants were recruited according to the established inclusion criteria. All the recruited patients agreed to participate and signed informed consent. Nine subjects (5F-4M) completed the study: their mean (SD) age was 44.22 (12.21), with a mean (SD) hours on wheelchair of 11.67 (2.45). The clinical characteristics can be synthetized as follow: mean (SD) years of injury 23.14 (12.21), lesion level C7 (1), L2 (1), L3 (1), T10 (1), T12 (2), T3 (1), T4 (2).

To assess that the rasterstereography was a reliable tool in evaluating the seated position of the person with spinal cord injury, the Cronbach's alpha was assessed and the tool was found to have an alpha of 0.74; intra-observer and inter-observer reliability were 0.91 (0.62–0.98) and 0.96 (0.88–0.99), respectively.

These values confirm the stability of the rastersteregraphy both in the evaluation of the same subject at different times by the same rater between t2 and t3 (intra-rater reliability) and in the measurements by three different operators assessed at t2 (inter-rater). To evaluate the validity of the rastersteregraphy, the Formetric 4D results were correlated with the anthropometric measurements recorded by the operator during the postural assessment. The correlations were evaluated using Spearman's rho; some statistically significant correlations were found, as reported in Table 1.

Table 1. Gold standard analysis: Spearman's rho correlation between Formetric 4D and anthropometric measures recorded through postural assessment procedure

	Pelvic width [Mm]	Pelvic length [Mm]	Floor-popliteal cable [Mm]	Trunk height [Mm]	SIAS distance [Mm]	SIPS distance [Mm]
PI (DL-DR) [°]	0.07	0.26	0.16	0.03	0.26	0.12
PIDLDR [mm]	0.11	0.33	0.20	0.09	0.24	0.04
Twist Emibacini DL-DR [°]	0.65	0.35	0.33	0.28	0.23	0.11
KA Kyphotic Apex (VPDM) [mm]	0.32	0.02	.78*	0.40	.73*	0.46
ITL inversion point [mm]	0.59	0.05	0.29	0.13	.77*	0.54
Lordotic apex LA (VPDM) [mm]	0.11	0.41	0.43	0.05	0.64	0.19
ILS reversal point [mm]	0.46	0.30	0.38	0.17	0.64	0.11
Cervical arrow (Stagnara) [mm]	0.44	−.73*	0.17	0.08	0.08	0.25
Lumbar arrow (Stagnara) [mm]	0.12	0.23	0.53	0.03	0.31	0.14
ICT-ITL angle (Max) [°]	0.18	0.47	0.25	0.11	0.34	0.13
ITL-ILS (Max)	0.17	0.37	0.07	0.60	0.15	0.18
Surface rotation (Rms) [°]	0.17	0.10	0.22	0.28	0.12	0.29
Surface rotation (Amplitude) [°]	0.23	0.20	0.11	0.10	0.38	0.04
VPDM lateral deviation (+Max) [mm]	0.38	0.31	0.44	0.40	0.21	0.22

(*continued*)

Table 1. (*continued*)

	Pelvic width [Mm]	Pelvic length [Mm]	Floor-popliteal cable [Mm]	Trunk height [Mm]	SIAS distance [Mm]	SIPS distance [Mm]
VPDM lateral deviation (−Max) [mm]	−.78*	−.79*	0.21	0.3	0.03	0.54
VPDM lateral deviation (Amplitude) [mm]	0.46	0.45	0.09	0.33	0.29	.81**

* p < 0.05; ** p < 0.01

Abbreviations: **VP** = prominent vertebra (spiny apophysis of C7); **DR and DL** = right and left lumbar dimple (Michaelis dimples); **DM** = average point of the segment that connects DR and DL; **ICT** = cervico-thoracic inversion (cervico-dorsal hinge); **ITL** = thoraco-lumbar inversion (back-lumbar hinge); **ILS** = lumbo-sacral inversion (lumbo-sacral hinge); **KA** = apex kyphosis (max curvature of the dorsal kyphosis); **LA** = apex lordosis (max curvature of lumbar lordosis); VP-DM trunk length: VP and DM joining segment length.; **DL-DR pelvic inclination**: vertical difference between DL and DR.; **Lateral deviation VP-DM**: on the frontal plane, horizontal lateral deviation of the centers of the vertebral bodies with respect to the joining line VP-DM (rms = quadratic mean, max = maximum value); **Lateral deviation VPDM (+max)**: max deviation to the right; **Lateral deviation VPDM (−max)**: max deviation to Sn; **Pelvic inclination (dimples)**: arithmetic mean of the 2 angles formed by the perpendicular to the surface in DR and DL and the vertical axis (mean pelvic torsion); **ICT-ITL (max) cytotic angle**: upper angle formed by tangents to the sagittal curve in ICT and ITL (represents the maximum value of the cytotic angle); **ITL-ILS lordotic angle (max)**: upper angle formed by tangents to the sagittal curve in ITL and ILS (represents the maximum value of lordotic angle); **SIAS distance [Mm]**: Antero-superior iliac spinal cord; **SIPS distance [Mm]** Posterior-superior iliac spine

4   Discussion

This study highlighted the ability of Formetric 4D to provide consistent results at different times and with different evaluators. Compared to the postural evaluation, the degree of accuracy of the measurements acquired with Formetric 4D, analyzed through Spearman's rho, showed positive correlations with anthropometric measurements, such as the width and length of the pelvis, the distance of the popliteal earth-hollow, trunk height, and the distance of the anterior-superior and posterior-superior iliac crests. These correlations are statistically significant, indicating a covariance with the following anatomical measurements: the torsion of emydines, the kyphotic apex, the lordotic apex, and the lordotic angle. However, these correlations currently do not allow for the precise definition of the elements that influence the proportional trend of the variables, but only show the covariance between the data. While other studies in the literature have shown the usability, sensitivity, and repeatability of the Formetric D with participants in an upright position, this pilot study demonstrated the reliability of the Formetric 4D as a tool for the follow-up and screening of this population in a sitting posture for the first time. This study also demonstrated how this tool can complement postural assessments done by occupational therapists, objectifying them. Expressing the anatomic characteristics, postural attitudes and the deformities, heretofore recordered subjectively by the therapist, in a concrete image. The correlation between these two different measurement instruments is an advantage not only at the clinical level, but also at the educational level. In fact, postural assessment is a very difficult assessment for students to learn, because it requires years of experience and specific anatomical knowledge. Accompanying the demonstrations of the practice with images collected technologically through rasterstereography, this method will be of great help in university teaching. Thanks to rasterstereography, in fact, students will be able to have a three-dimensional view of the shape of the back, the spine, and the position of the pelvis in a sitting position and can compare it with the objective examination.

The limited number of samples although it was defined in relation to the type of study can represent a limitation of the study. A larger sample of the population would have allowed greater generalization of the results. Therefore, for future studies on the topic, we hope to increase the sample selection; this could, for example, collect information about subjects with different diagnoses, allowing additional comparisons between the measured variables, and taking into account differences in the population.

5 Conclusions

The present study provides information enabling the use of the Formetric 4D tool in clinical, research, and educational settings. In the clinical field, the tool will be useful for occupational therapists in analyzing the postural characteristics of the spine and the pelvis, providing an objectified postural evaluation. This allows therapists to obtain objectively valid data that will be useful for preventing deformities and carrying out postural follow-ups. In the field of research, the Formetric 4D will be useful as it provides a starting point for expanding the research for wheelchair posture broadening the research to other diseases or using different support surfaces. And finally, at the educational level, this will be a very useful tool that allows students to have a three-dimensional representation of the anatomical components involved in the sitting position, helping them to learn and gain an in-depth understanding of how to perform an objective postural assessment examination.

Funding. None.

Compliance with Ethical Standards. We certify that all applicable institutional and governmental regulations concerning the ethical use of human volunteers were followed during the course of this research.

Conflict of Interest. All authors declare no conflict of interest.

References

1. Sonenblum, S.E., Sprigle, S.H., Martin, J.S.: Everyday sitting behavior of full-time wheelchair users. J. Rehabil. Res. Dev. **53**, 585–598 (2016)
2. Mattie, J., Borisoff, J., Miller, W.C., Noureddin, B.: Characterizing the community use of an ultralight wheelchair with "on the fly" adjustable seating functions: a pilot study. PLoS One **12**, e0173662 (2017)
3. International Organization for Standardization. ISO 16840-1:2013 Wheelchair seating – Part 1: Vocabulary, reference axis convention and measures for body segments, posture and postural support surfaces [Internet]. 2nd edn (2013). https://www.iso.org/standard/42064.html
4. International Organization for Standardization. ISO 16840-2:2018 Wheelchair seating – Part 2: Determination of physical and mechanical characteristics of seat cushions intended to manage tissue integrity [Internet]. 2nd edn (2018). https://www.iso.org/standard/66972.html
5. Cohen, L.J., Fitzgerald, S.G., Lane, S., Boninger, M.L., Minkel, J., McCue, M.: Validation of the seating and mobility script concordance test. Assist. Technol. **21**, 47–56 (2009)
6. Tabard-Fougère, A., Bonnefoy-Mazure, A., Hanquinet, S., Lascombes, P., Armand, S., Dayer, R.: Validity and reliability of spine rasterstereography in patients with adolescent idiopathic scoliosis. Spine **42**, 98–105 (2017)
7. Scheidt, S., Endreß, S., Gesicki, M., Hofmann, U.K.: Using video rasterstereography and treadmill gait analysis as a tool for evaluating postoperative outcome after lumbar spinal fusion. Gait Posture **64**, 18–24 (2018)
8. Padulo, J., Ardigò, L.P.: Formetric 4D rasterstereography. BioMed Res. Int. (2014)
9. Guidetti, L., Bonavolontà, V., Tito, A., Reis, V.M., Gallotta, M.C., Baldari, C.: Intra- and interday reliability of spine rasterstereography. Biomed. Res. Int. (2013)

Effectiveness of SaeboReJoyce in the Evaluation of the Improvement of the Occupational Performance in Parkinson's Disease: An Outcome Research

Silvia Salviani[1], Marco Tofani[2], Giovanni Fabbrini[3], Antonio Leo[4], Anna Berardi[1], Julita Sansoni[5], and Giovanni Galeoto[5(✉)]

[1] Sapienza Università di Roma, Rome, Italy
[2] Neurorehabilitation Unit, Bambino Gesù Children's Hospital, Rome, Italy
[3] Department of Human Neurosciences, Sapienza University of Rome, Rome, Italy
[4] Officine Ortopediche Rizzoli Srl, Rome, Italy
[5] Department of Public Health and Infectious Diseases, Sapienza University of Rome, Rome, Italy
giovanni.galeoto@uniroma1.it

Abstract. The aim of this study was to assess the effectiveness of SaeboReJoyce in the evaluation of improvement of the occupational performance, in Parkinson's disease. Six patients with Parkinson's disease were recruited between June 2018 and September 2018 from the Parkinson's disease Outpatient Clinic of "Policlinico Umberto I" hospital in Rome. We assessed the effectiveness of SaeboReJoyce in evaluating the improvement of the occupational performance in Activities of Daily Living and Instrumental Activities of Daily Living. We chose four different activities that patients performed for five weeks with two weekly sessions. The impact on the quality of life, movements, fine and gross grasps, depression and satisfaction were evaluated. The analysis of the data from the rating scales and the tests shows an improvement in the functionality of the upper limb. The results for the rating scales are not statically significant but clinically relevant, while results for the ReJoyce are in part statistically significant. Our results suggest that the SaeboReJoyce can be a useful instrument not only for the health professional dealing with Parkinson Disease but also for nursing and rehabilitation sciences students for the evaluation of changes in functionality of the upper limb, including both gross and fine motor tasks/grasps, for patients with Parkinson Disease.

Keywords: SaeboReJoyce · ADL · IADL · Parkinson's disease · Upper limb

1 Introduction

Parkinson's disease (PD) is a common progressive, neurodegenerative disease, affecting 0.1% of the general population and 1% of the population over 65 years [1]. In 2016, the estimated regional incidence rate of PD in Italy was 0.28 new cases/1000 person-years, with a prevalence of 3.89/1000 persons [2]. The Global Burden of

---

The original version of this chapter was revised: The author names have been updated. The correction to this chapter is available at https://doi.org/10.1007/978-3-030-23884-1_20

E. Popescu et al. (Eds.): MIS4TEL 2019, AISC 1008, pp. 63–70, 2020.
https://doi.org/10.1007/978-3-030-23884-1_9

Disease Study estimates that by 2040 as many as 12.9 million individuals will be affected, due to the aging of the world's population, and keeping in consideration that the incidence of PD increases with age [3].

PD has relevant clinical, social and economic implications. Disability in PD is due to the presence of both motor and non-motor symptoms, which restrict both the self-sufficiency and social participation of patients, leading to a low quality of life (QoL) for both patients and their caregivers [4].

The rating scales and standardized treatments currently available in Occupational Therapy are insufficient, thus leading to a difficulty in both students training and drafting a correct intervention strategy for the patients.

The SaeboReJoyce [5] is a rehabilitation device that allows to measure and evaluate the range of movement of the upper limb and different types of grasps. It has been used in tetraplegic patients jointly with the functional electrical stimulation and the conventional exercise therapy [6, 7]. The SaeboReJoyce offers an additional tool to face this problem, being a novel rehabilitation device designed to measure, evaluate and increase the strength of muscles and the range of motion of joints of the upper extremities [5].

To date, this tool has been used with C5–C7 tetraplegic patients to play computer games associated with Activities of Daily Living (ADLs) [6, 7].

The aim of this study was to evaluate its effectiveness in the evaluation of the improvement of occupational performance and functionality of the upper limb in ADLs and Instrumental Activity of Daily Living (IADLs), in PD.

2   Material and Methods

2.1   Population

The present study was conducted by health professionals of Sapienza University of Rome and ROMA – Rehabilitation & Outcome Measures Assessment Association.

We included patients from the Parkinson's disease Outpatient clinic of "Policlinico Umberto I" hospital in Rome from June 2018 to September 2018. The inclusion criteria were diagnosis of Parkinson's disease, age > 40 years, Montreal Cognitive Assessment MoCA [8] score $\geq$ 26 and knowledge of the Italian language. For detecting mild cognitive impairment during patients' selection, we used the MoCA, a 30-point test administered in 10 min (cut off" 26) [8].

2.2   Rating Scale and Tests

To evaluate QoL we used a patient-assessed instrument, the **Parkinson's Disease Questionnaire-39** (PDQ-39) the most thoroughly tested and applied assessment tool in PD patients. It consists of 39 items, divided into 8 subtests: mobility, activities of daily living, emotional wellbeing, stigma, social support, cognition, communication and bodily discomfort (0 = not disability; 156 = severe disability) [9]. We use the **Disability of the Arm, Shoulder and Hand** (DASH) to identify symptoms and functions of the upper limb in patients with neurological or orthopedic disorders and it consists of

30 questions (0 = not disability; 100 = severe disability) [10]. The **Jebsen-Taylor Hand Function Test** (JTHFT) is a standardized test to evaluate the upper limb functionality through 7 items that simulated ADL [11] - The **ReJoyce Arm-Hand Function Test** (RAHFT) reproduces every movement and gross and fine motor tasks/grasps possible with the SaeboReJoyce [4, 5]. We performed three evaluations, during the first (T_0), the fifth (T_1) and the last (T_2) session. At last, we used the **Geriatric Depression Scale** (GDS) to evaluate the level of depressive symptoms over the past week considers the patient's point of view. It is useful for the screening and measurement of the degree of perceived depression in patients with PD and it is composed by 30 items [12, 13].

2.3   Follow-Up Evaluation

As a preliminary step we carried out an interview with each participant during which every person was specifically informed about the procedures and aims of the study; we collected personal data necessary for the research after the acquisition of informed consent from each participant. The evaluations were performed using self-administrated measurement tools during the first (T0) and the last (T1) session. Only the ReJoyce Arm-Hand Function Test was performed three times, during the first (T0), the fifth (T1) and the last (T2) session.

2.4   Intervention

The treatment lasted five weeks with two weekly sessions lasting about 45 min, except the first and the last meeting. In fact, during the first session, we outlined the project, obtained informed consent and in both the first and last sessions. The evaluations were performed using the measurement tools.

The SaeboReJoyce System is made up of 4 parts: the base, that clamps to a desk; the arm supports the manipulandum and provides resistance to motion in all directions; the manipulandum that consists of the peg, coins, jar lid, doorknob, key, gripper and handles used to play the computer games; and the PC Laptop that runs the interactive SaeboReJoyce NeuroGaming™ software.

The 4 chosen activities included: Bartender, Catcher, Weedo and Stack Attack. The goal of Barthender is to renovate the bar by pouring beverages for patrons based on their orders as fast and accurately as possible. The player has to select the beverage moving the manipulandum to the top of the screen over the requested bottle and grasp it. Once come back to the counter-top, a glass will appear and the player must "swing the gripper to pour until the golden marker". The goal of Catcher is to catch as many of the falling color-coded objects in the color-coded buckets as possible. We chose the jar lid like control. The player of Weedo must protect the flowers by removing weeds and dropping them into the compost heaps situated on either side of the screen. Like control, we chose the peg. The player of Stack Attack must eliminate falling blocks as they pass through bumpersby performing the displayed hand function. The game is over if too many blocks stack at the bottom of the screen. The preset hand functions are: gripper, peg, doorknob and jar lid in either directions [5]. The 4 chosen activities are illustrated in Fig. 1a, b, c and d.

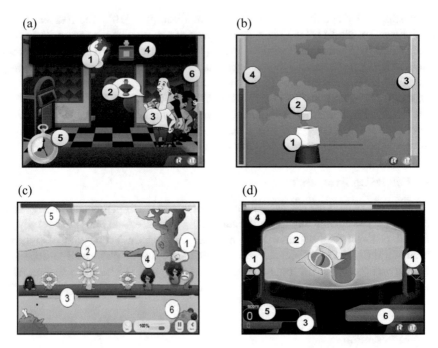

Fig. 1. (a) The interactive scenarios of Bartender. (b) The interactive scenarios of Catcher. (c) The interactive scenarios of Wedoo. (d) The interactive scenarios of Stack Attack.

2.5 Analysis of Data

We used the descriptive analysis of general information (age, sex, diagnosis, year of diagnosis, Hoehn&Yahr stage, education, employment, marital status).

Concerning the SaeboReJoyce test results, we grouped data in 3 groups, by movement (left, right, up, down, forwards and backwards), rotational movement (doorknob, key and jar) and gross and fine grasps (grip, peg and coin). For each group, the data were analysed both in terms of time required to complete the action (time) and quality of execution (value).

In order to provide a reliable evaluation of the change in upper limb functionality, we used the non-parametric Wilcoxon test (Z). The significance level has been set for p-value less than or equal to 0.05. The statistical analysis was performed through the Statistical Package of Social Science (SPSS) – version 23.0 for Windows.

3 Results

Fourteen patients were recruited for this study between June 2018 and September 2018. They all signed the written informed consent [14, 15] and six patients completed the study. Mean age of participants was 65.38 ± 5.26 SD and the mean MoCA score was

26.75 ± 0.89 SD. The features of the sample are presented in Table 1. The results of the rating scales and of the Jebsen test are reported respectively in Tables 2 and 3. The statistically significant data of the SaeboReJoyce test are summarized in Table 4.

Table 1. Features of the 6 participants.

	Sample = 6
Gender male N (%)	5 (83,3)
Age mean (SD)	65.50 ± 5.92
Years from diagnosis Mean (SD)	6.33 ± 5.39
MoCA Mean (SD)	26.83 ± 0.98
H&Y N (%)	
1	1 (16.7)
2	2 (33.3)
3	3 (50)
Education N (%)	
Elementary School	2 (33.3)
Middle School	1 (16.7)
High School	3 (50)
Employment N (%)	
Employed	2 (33.3)
Unemployed	1 (16.7)
Retired	3 (50)
Marital Status N (%)	
Married	5 (62.5)
Divorced/Separated	1 (16.7)

Table 2. Z-test results of the rating scales PDQ-39, DASH and GDS.

	T_0 μ ± SD	T_1 μ ± SD	Test Z	p
PDQ-39	84.00 ± 32.00	77.00 ± 26.28	−.674	0.500
DASH	62.88 ± 32.63	55.33 ± 22.18	−.530	0.596
GDS	11.63 ± 6.30	10.33 ± 8.14	−.135	0.892

Table 3. Z-test results of the Jebsen test; DH = dominant hand; NDH = non-dominant hand.

	T_0 μ ± SD	T_1 μ ± SD	Test Z	p
Item 1 DH	48.00 ± 56.80	27.38 ± 22.84	−.734	0.463
Item 2 DH	12.47 ± 9.77	7.14 ± 3.67	−.943	0.345
Item 3 DH	12.22 ± 8.85	8.14 ± 2.17	−.943	0.345
Item 4 DH	31.95 ± 59.88	8.65 ± 2.37	−1.782	0.075
Item 5 DH	10.44 ± 4.57	8.33 ± 4.49	−1.363	0.173

(*continued*)

Table 3. (*continued*)

	T_0 μ ± SD	T_1 μ ± SD	Test Z	p
Item 6 DH	10.56 ± 11.49	5.67 ± 1.87	−.674	0.500
Item 7 DH	27.71 ± 61.55	5.57 ± 2.18	−.105	0.917
Item 1 NDH	68.12 ± 29.02	60.36 ± 22.93	−.734	0.463
Item 2 NDH	9.66 ± 2.72	7.84 ± 2.55	−1.363	0.173
Item 3 NDH	9.76 ± 2.26	9.63 ± 2.25	−.105	0.917
Item 4 NDH	13.85 ± 3.14	12.48 ± 3.78	−.105	0.917
Item 5 NDH	10.10 ± 3.78	8.62 ± 3.16	−1.572	0.116
Item 6 NDH	7.21 ± 2.37	6.41 ± 2.28	−.524	0.600
Item 7 NDH	6.86 ± 2.10	5.92 ± 2.15	−.314	0.753

Table 4. Z-test results of the ReJoyce test.

	T_0 μ (SD)	T_1 μ (SD)	T_2μ (SD)	T_0–T_1 p value	T_0–T_2 p value
Movement to left - RH - value	64 (25.5)	93.3 (11.5)	98.6 (2.2)	0.043*	0.043*
Movement to right - RH - value	57.4 (25.2)	82.8 (22.1)	86 (12.8)	0.068	0.043*
Gross grasp - movement - release - RH - value	61 (27.3)	88.7 (3.7)	90.7 (2.9)	0.028*	0.027*
Gross grasp - release - RH - value	81.7(25.75)	95(5.5)	97 (7.4)	0.500	0.042*
Fine grasp - movement - release - RH – value	59.7 (28.2)	89.7 (3)	88.7 (3.9)	0.027*	0.028*
Gross grasp - movement - release - LH - value	67.1 (29.6)	86.3 (6)	90 (4.2)	0.058	0.028*
Fine grasp - movement - release - LH - value	72.4 (13.1)	88.5 (1.8)	88.5 (2.9)	0.027*	0.027*
Movement to left - RH - time	92.4 (5)	95.3 (6.7)	96.3 (2.3)	0.173	0.046*
Movement to right - RH - time	90.4 (7.6)	95.7 (6.3)	93.7 (4.5)	0.046*	0.225
Movement up - RH - time	92.1 (5)	97.5 (2.2)	97 (3.2)	0.046*	0.027*
Movement down - RH - time	88.2 (9.7)	94.2 (4.5)	95.2 (2.5)	0.080	0.043*
Movement backwards - RH - time	94.3 (4.7)	96.5 (3.6)	98 (2.4)	0.131	0.027*
Movement to left - LH - time	90.1 (8.7)	94.2 (4.8)	95.5 (3.5)	0.046*	0.080
Movement up - LH - time	92.3 (6.5)	96.3 (3.6)	95.3 (3.8)	0.046*	0.144
Gross grasp - movement - release - RH - time	61 (27.3)	88.7 (3.7)	90.7 (2.9)	0.028*	0.027*
Fine grasp - movement - release - RH - time	59.7 (28.7)	89.7 (3)	88.7 (3.9)	0.027*	0.028*
Gross grasp - movement - release - LH - time	67.1 (29.6)	86.3 (6)	90 (4.2)	0.058	0.028*
Fine grasp - movement - release - LH - time	72.4 (13.1)	88.5 (1.8)	88.5 (2.9)	0.027*	0.027*

*p < 0.05

4   Discussion

The aim of this study was to widen Occupational Therapist's standardized and validated instruments through the assessment of the efficacy of the SaeboReJoyce in the evaluation of changes in mobility and functionality of the upper limb in patients with PD.

Even if the evidence has not yet resulted in standardized guidelines, Virtual reality (VR) has recently been used as a tool for rehabilitation in individuals with different neurological disease, included PD. VR is able to facilitate motor learning for balance and gait rehabilitation resulting in additional benefits, especially when combined with other interventions such as conventional rehabilitation [16].

Thanks to the use of immersive VR, stress decreased, the level of arousal after exposure increased and there were not negative physiological and psychological effects [17].

Concerning the rating scale (PDQ-39, DASH, GDS and to the Jebsen test), a statistically significant difference not in all the items of the ReJoyce was found, however it is considered clinically relevant. For most items of the SaeboReJoyce a statistical difference was found with a $p < 0.05$.

4.1 Limitations of the Study

Our study has two important limitations: the small sample size and the short duration of treatment.

5 Conclusion

Our data suggest that the SaeboReJoyce could be an useful instrument not only for the health professional dealing with PD but also for students for the evaluation of changes in functionality of the upper limb, including both gross and fine motor tasks/grasps, for patients with PD. It could be also useful for the health professional dealing with PD, allowing them to elaborate a correct intervention strategy and modify it if necessary.

In order to better evaluate the effectiveness of SaeboReJoyce as standardized instrument for the evaluation of upper limb before, during and after rehabilitation, further studies.

References

1. De Lau, L.M.L., Breteler, M.M.B.: Epidemiology of Parkinson's disease. Lancet Neurol. **5**(6), 525–535 (2006)
2. Valent, F., Devigili, G., Rinaldo, S., Del Zotto, S., Tullio, A., Eleopra, R.: The epidemiology of Parkinson's disease in the Italian region Friuli Venezia Giulia: a population-based study with administrative data. Neurol. Sci. **39**(4), 699–704 (2018)
3. Dorsey, E.R., Bloem, B.R.: The Parkinson pandemic – a call to action. JAMA Neurol. **75**(1), 9–10 (2018)
4. Alves, G., Forsaa, E.B., Pedersen, K.F., Dreetz Gjerstad, M., Larsen, J.P.: Epidemiology of Parkinson's disease. J. Neurol. **255**(5), 18–32 (2008)
5. Clinical presentation. https://www.saebo.com/saebo-rejoyce/. Accessed 11 Jan 2019
6. Kowalczewski, J., Chong, S.L., Galea, M., Prochazka, A.: In-home tele-rehabilitation improves tetraplegic hand function. Neurorehabil. Neural Repair **25**, 412–422 (2011)
7. Kowalczewski, J., Ravid, E., Prochazka, A.: Fully-automated test of upper-extremity function. In: 33rd Annual International Conference of the IEEE EMBS Boston, Massachusetts USA, 30 August–3 September 2011

8. Nasreddine, Z.S., Phillips, N.A., Bédirian, V., Charbonneau, S., Whitehead, V., Colli, I., Cummings, J.L., Chertkow, H.: The Montreal cognitive assessment, MoCA: a brief screening tool for mild cognitive impairment. J. Am. Geriatr. Soc. **53**, 695–699 (2005)

9. Galeoto, G., Colalelli, F., Massai, P., Berardi, A., Tofani, M., Pierantozzi, M., Servadio, A., Fabbrini, A., Fabbrini, G.: Quality of life in Parkinson's disease: Italian validation of the Parkinson's Disease Questionnaire (PDQ-39-IT). Neurol. Sci. **39**, 1903–1909 (2018)

10. Padua, R., Padua, L., Ceccarelli, E., Romanini, E., Zanoli, G., Amadio, P.C., Campi, A.: Italian version of the disability of the arm, shoulder and hand (DASH) questionnaire. Cross-cultural adaptation and validation. J. Hand Surg. (Br. Eur. Volume) **28B**(2), 179–186 (2003)

11. Culicchia, G., Nobilia, M., Asturi, M., Santilli, V., Paoloni, M., De Santis, R., Galeoto, G.: Cross-cultural adaptation and validation of the Jebsen-Taylor hand function test in an Italian population. Rehabil. Res. Pract. (2016)

12. Massai, P., Colalelli, F., Sansoni, J., Valente, D., Tofani, M., Fabbrini, G., Fabbrini, A., Scuccimarri, M., Galeoto, G.: Reliability and validity of the geriatric depression scale in Italian subjects with Parkinson's disease. Parkinson's Dis. (2018)

13. Galeoto, G., Sansoni, J., Scuccimarri, M., Bruni, V., De Santis, R., Colucci, M., Valente, D., Tofani, M.: A psychometric properties evaluation of the Italian version of the geriatric depression scale. Depression Res. Treat. (2018)

14. Galeoto, G., De Santis, R., Marcolini, A., Cinelli, A., Cecchi, R.: IL consenso informato in Terapia Occupazionale: proposta di una modulistica. G. Ital. Med. Lav. Ergon. **38**, 107–115 (2016)

15. Galeoto, G., Mollica, R., Astorino, O., Cecchi, R.: Informed consent in physiotherapy: proposal of a form. G. Ital. Med. Lav. Ergon. **37**, 245–254 (2015)

16. Cano Porras, D., Siesmonsma, P., Inzelberg, R., Zelig, G., Plotnik, M.: Advantages of virtual reality in the rehabilitation of balance and gait: systematic review. Neurology **90**, 1017–1025 (2018)

17. Kim, A., Darakjian, N., Finley, J.M.: Walking in fully immersive virtual environments: an evaluation of potential adverse effects in older adults and individuals with Parkinson's disease. J. NeuroEng. Rehabil. **14**, 16 (2017)

High-Fidelity Simulation Type Technique Efficient for Learning Nursing Disciplines in the Courses of Study: An Integrative Review

Ilaria Farina, Albina Paterniani, Giovanni Galeoto, Milena Sorrentino, AnnaRita Marucci, and Julita Sansoni[✉]

Nursing Research Unit - Public Health and Infectious Disease, Sapienza University of Rome, Piazzale Aldo Moro 5, 00185 Rome, Italy
julita.sansoni@uniromal.it

Abstract. Background: Simulation is considered an effective strategy for educating nursing students and practicing nurses in several clinical settings. We can therefore define the simulation as a technique to replace or report real cases with guided experiences that replicate fundamental aspects of the welfare practice in an interactive way. The aim of this study is to determine which of the different types of high-fidelity simulation is most effective as a teaching method for nurse students. Research design: A integrative review of the literature was undertaken following the framework of Whittemore and Knafl (2005); an electronic search of literature was conducted using 3 databases (CINAHL, SCOPUS and MEDLINE). This final screening yielded 18 articles, after the exclusion of articles consisting of only abstracts, review articles, and articles not relevant to the topic. Discussion: The integrative review in high-fidelity simulation type show that knowledge achieved via simulation is remembered for a longer time respect to knowledge achieved via lecture. After analyzing the data, 3 main themes were identified. Conclusion: Engaging in a realistic situation, as any type of high-fidelity simulation, provides the context for nursing students, who are often concrete thinkers, to expand their clinical reasoning and "sense of salience". Application of knowledge is essential to insure safe, effective practice in today's challenging healthcare setting.

Keywords: Nursing student · Post graduate nursing student · High fidelity simulation

1 Introduction

The patient death caused by a medical error has the third highest overall mortality rate in US, and an annual death rate of 134,581 hospitalized patients was recently reported [1]. The growing attention to patient safety has shown that in Italy 78% of errors are attributable on average to "Human Factor" and, unfortunately, are a recognized phenomenon in health care [2]. The provision of unsafe medical services affects not only patients' lives by causing physical injury [3] but also leads to emotional stress, depression, and guilt in medical professionals [4]. The "Human factor" is a training that, as the term means, studies the "human factor", the interactions of man with other

© Springer Nature Switzerland AG 2020
E. Popescu et al. (Eds.): MIS4TEL 2019, AISC 1008, pp. 71–76, 2020.
https://doi.org/10.1007/978-3-030-23884-1_10

individuals, systems, equipment and machines during his work to correct certain risk connected behaviours. The training objective is to increase the levels of safety of interventions and overall efficiency. The aviation industry has adopted the peculiar education pipelines on the basis of these evidences for decades: learning techniques and procedures through the use of simulation and the principles of Crew/Crisis Resource Management (CRM) [5]. The most important aspect is that the simulation makes it possible to establish one or more health care standards aimed to a specific patient/user population, according to which all the nursing services provided can be measured and standardized [6]. We can therefore define the simulation as a technique to replace or report real cases with guided experiences that replicate fundamental aspects of the welfare practice in an interactive way, giving it numerous advantages such as: replicability, personalization of learning on the single professional, and learning linked to repetition and to prevent errors. Simulation in nursing education can be categorized into three main types – low fidelity, medium fidelity, and high fidelity. High fidelity simulation is defined as a teaching method which produces realistic clinical situations in a protected environment [7]. The fidelity or credibility of the situation is an important element to consider when evaluating the meaning of the fullness of the simulation.

The aim of this work is to determine which of the different types of high-fidelity simulation is the most effective as teaching method for nursing students and the type of impact on clinical learning.

2   Materials and Method

Research Design: An integrative review of the literature follow the steps outlined by Whittemore and Knafl [8]. This framework describes a review of the integrative literature as a research that can integrate qualitative and quantitative studies with mixed method studies. In this paper, we used the method described above that allowed a wider selection of articles to be included in the study, providing an overview of high-fidelity simulation types and their impact on the effectiveness of clinical nursing learning. An electronic search of literature was conducted using 3 databases (CINAHL, SCOPUS and MEDLINE). Additional literature was obtained by systematic checking the reference lists of all identified articles. Search terms included "simulation", "nursing", "high fidelity simulation", "nursing student", and "postgraduate nursing student" keywords. Medical subject headings were used in "advanced searches" of the 3 used international databases. Keywords were combined through the use of Boolean operators such as "OR", "AND". Articles were chosen based on the following criteria: (1) articles that measured the knowledge and/or attitudes of nurses who had performed the simulation, (2) full-text and peer-reviewed studies, (3) both qualitative and quantitative research articles, (4) articles published in the English language, and (5) from 2014 to 2019. In addition, editorials, letters, books were excluded from the study. Studies were appraised using a critical appraisal tool: the Critical Appraisal Skills Programme tool (CASP 2017) [9] for qualitative studies and the analysis technique Systematic reviews. Center for Reviews and Dissemination (CRD) guidance for undertaking reviews in health care CRD [10] was used to analyze quantitative and mixed method study. These two tools have been used to ensure the validity of the instrument used and the justification of the methodological approach of the study.

3 Result

The research trough published literature yielded 630 titles for review. The titles and abstracts of these articles were screened according to the inclusion criteria reported in the previous section. This final screening yielded 18 articles, after the exclusion on articles consisting of only abstracts and articles not relevant to the topic. Figure 1 is a flow diagram used during the literature screening and selection.

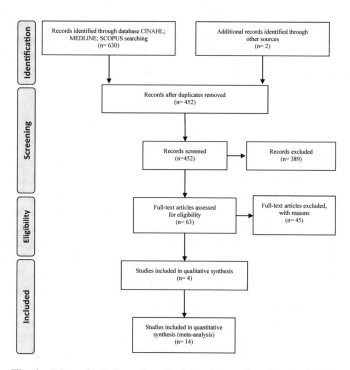

Fig. 1. Prisma 2009 flow diagram about the conducted analysis [11].

Following the Whittemore and Kraft [8] methodology, each study has been read several times and the citations of the results have been reported in a table and ordered according to the year of publication, specifying in the table the type of study management method. Qualitative and quantitative results were integrated as foreseen by the reference framework [8]. Results similar to all studies were organized and grouped during the comparison phase using thematic analyzing method [12].

Data extraction was done by two researchers independently and reviewed by a third researcher whenever necessary. A summary of research information (table available on request) was prepared including authors' names, country, high-fidelity simulation type, major findings and study type.

A percentage of 60% of the studies used a quantitative method, 4 studies used a qualitative research design and 2 studies used a mixed method. Sample sizes ranged from 23 to 390 nurses. Convenient sampling was the main method of participant recruitment used. Research Setting: The studies were from across the globe, but the largest number per country (n = 8) was from the United States. The study participants were nursing student and post-graduate nursing student. Research method of high-fidelity simulation: 6 of the 18 included articles used the high-fidelity simulation technique with patient simulator. 14 articles dealt with simulation using realistic mannequins. After analyzing the data, 3 main themes were identified; these are the critical clinical judgment, the competence of transversal communication and empathic realism. The topic found in 9 articles [13–21] is the critical clinical judgment, followed by empathic realism (6 articles) [21–26] and transversal communication competence (5 articles) [23, 25–28] in descending order. As reported, there is not a great difference in the presence of themes, and part of the studies has reported a themes combination.

4   Discussion

Integrative reviews are useful educational articles because of put many pieces of information together into a readable format. They are helpful in presenting a broad perspective on a topic and often describe the history or development of a problem, or its management. The integrative review in high-fidelity simulation type show that knowledge achieved via simulation is remembered for a longer time respect to knowledge achieved via lecture [23]. Simulation is a tool that increases exposure to events in a safe and supportive environment, allowing trainees to develop skills without harming patients and without penalties for the trainees themselves [23]. The history of simulation in medical education and possible future directions. Simulation scenarios can be tailored as needed, allowing for customized trainee education. This aspect has been included in the topic "critical clinical judgment" [13–21] found in the majority of the selected articles. The nursing students exposed to every type of high-fidelity simulation can think in a critical way about the simulation performance. Critical reflection on clinical judgment creates in students a feeling of self-control that helps them gain confidence in themselves. The second theme that emerged from the integration of the articles was the "cross-communication competence" [23, 25–28] capacity acquired by the students following the high-fidelity simulation scenario where more than one healthcare professional interacted and where the standardized patient was used.

The last theme that emerged is "empathy" [21, 22, 24–27] this sentiment came from the realism of various types of high fidelity simulations. In literature there is unequivocal evidence of the physiological and psychosocial impact that gives empathic commitment to patients especially for the nursing profession. Specifically, the simulation studies included in this work show that the direct interaction with standardized patients (SPs) has positive effects not only on nursing competence but also on their problem-solving ability and communication competence. The studies related to high fidelity simulation with the use of mannequin show that nurse student developed their skills and competences by working with other students as well as through the feedback and reflections given after the high-fidelity simulation.

5   Conclusion

Nursing education must assure that the different patient care situations (i.e. case study, simulation with computerized manikins, problem-based learning) are effectively used in classroom settings according to the Whittemore and Kraft methodology [8]. Engaging in a realistic situation, as any type of high-fidelity simulation, the nursing students, who are often concrete thinkers, can expand their clinical reasoning and "sense of salience". Application of knowledge is essential to insure safe, effective practice in today's challenging healthcare setting. Improvement in confidence and clinical performance has the potential to influence not only the quality of patient care through competent practice, but also to influence staff engagement, retention, and the overall patient experience.

References

1. Makary, M.A., Daniel, M.: Medical error-the third leading cause of death in the US. BMJ **353**, i2139 (2016)
2. Focarelli, D.: La responsabilità sanitaria: problemi e prospettive (2015). www.ania.it/export/sites/default/it/pubblicazioni/monografie-e-interventi/Danni/Responsabilita-sanitaria-e-assicurazioni-proposte-e-criticita-Intervento-Focarelli-19.03.2015.pdf%0D
3. Berwick, D.M., Hackbarth, A.D.: Eliminating waste in US health care. JAMA, J. Am. Med. Assoc. **307**(14), 1513–1516 (2012)
4. Hobgood, C., Hevia, A., Tamayo-Sarver, J.H., Weiner, B., Riviello, R.: The influence of the causes and contexts of medical errors on emergency medicine residents' responses to their errors: an exploration. Acad. Med. **80**(8), 758–764 (2005)
5. Barton, G., Bruce, A., Schreiber, R.: Teaching nurses teamwork: Integrative review of competency-based team training in nursing education. Nurse Educ. Pract. **32**, 129–137 (2018)
6. La Cerra, C., Dante, A., Caponnetto, V., Franconi, I., Gaxhja, E., Petrucci, C., Alfes, C.M., Lancia, L.: Effects of high-fidelity simulation based on life-threatening clinical condition scenarios on learning outcomes of undergraduate and postgraduate nursing students: a systematic review and meta-analysis. BMJ Open **9**(2), e025306 (2019)
7. Leigh, G.T.: High-fidelity patient simulation and nursing students' self-efficacy: a review of the literature. Int. J. Nurs. Educ. Sch.
8. Whittemore, R., Knafl, K.: The integrative review: updated methodology. J. Adv. Nurs. **52**(5), 546–553 (2005)
9. Critical Appraisal Skills Programme (CASP): CASP checklists, Oxford (2013)
10. Khan, K.S., Ter Riet, G., Glanville, J., Sowden, A.J., Kleijnen, J.: Undertaking systematic reviews of research on effectiveness: CRD's guidance for those carrying out or commissioning reviews, 2nd edn, no. 4. CRD Report (2001)
11. Moher, D., Liberati, A., Tetzlaff, J., Altman, D.G.: Preferred reporting items for systematic reviews and meta-analyses: the PRISMA statement. PLoS Med. **6**(7), e1000097 (2009)

12. Braun, V., Clarke, V.: Using thematic analysis in psychology. Qual. Res. Psychol. **3**(2), 77–101 (2006)

13. Nevin, M., Neill, F., Mulkerrins, J.: Preparing the nursing student for internship in a pre-registration nursing program: developing a problem based approach with the use of high fidelity simulation equipment. Nurse Educ. Pract. **14**(2), 154–159 (2014)

14. Ignacio, J., Dolmans, D., Scherpbier, A., Rethans, J.J., Chan, S., Liaw, S.Y.: Comparison of standardized patients with high-fidelity simulators for managing stress and improving performance in clinical deterioration: a mixed methods study. Nurse Educ. Today **35**(12), 1161–1168 (2015)

15. Lee, J., Lee, Y., Lee, S., Bae, J.: Effects of high-fidelity patient simulation led clinical reasoning course: focused on nursing core competencies, problem solving, and academic self-efficacy. Japan J Nurs. Sci. **13**(1), 20–28 (2016)

16. Najjar, R.H., Lyman, B., Miehl, N.: Nursing students' experiences with high-fidelity simulation. Int. J. Nurs. Educ. Sch. **12**(1), 27–35 (2015)

17. Basak, T., Unver, V., Moss, J., Watts, P., Gaioso, V.: Beginning and advanced students' perceptions of the use of low- and high-fidelity mannequins in nursing simulation. Nurse Educ. Today **36**, 37–43 (2016)

18. Kable, A.K., Levett-Jones, T.L., Arthur, C., Reid-Searl, K., Humphreys, M., Morris, S., et al.: A cross-national study to objectively evaluate the quality of diverse simulation approaches for undergraduate nursing students. Nurse Educ. Pract. **28**, 248–256 (2018)

19. Vermeulen, J., Beeckman, K., Turcksin, R., Van Winkel, L., Gucciardo, L., Laubach, M., et al.: The experiences of last-year student midwives with high-fidelity perinatal simulation training: a qualitative descriptive study. Women Birth **30**(3), 253–261 (2017)

20. Unver, V., Basak, T., Tastan, S., Kok, G., Guvenc, G., Demirtas, A., et al.: Analysis of the effects of high-fidelity simulation on nursing students' perceptions of their preparedness for disasters. Int. Emerg. Nurs. **38**, 3–9 (2018)

21. Expósito, J.S., Costa, C.L., Agea, J.L.D., Izquierdo, M.D.C., Rodríguez, D.J.: Socio-emotional competencies as predictors of performance of nursing students in simulated clinical practice. Nurse Educ. Pract. **32**, 122–128 (2018)

22. Ward, J.: The empathy enigma: does it still exist? Comparison of empathy using students and standardized actors. Nurse Educ. **41**(3), 134–138 (2016)

23. Dunnington, R.M.: The centricity of presence in scenario-based high fidelity human patient simulation: a model. Nurs. Sci. Q. **28**(1), 64–73 (2015)

24. Bradley, P.: The history of simulation in medical education and possible future directions. Med. Educ. **40**(3), 254–262 (2006)

25. Kirkman, T., Hall, C., Winston, R., Pierce, V.: Strategies for implementing a multiple patient simulation scenario. Nurse Educ. Today **64**, 11–15 (2018)

26. Fusco, N.M., Foltz-Ramos, K.: Measuring changes in pharmacy and nursing students' perceptions following an interprofessional high-fidelity simulation experience. J. Interprofessional Care **32**(5), 648–652 (2018)

27. Bolesta, S., Chmil, J.V.: Interprofessional education among student health professionals using human patient simulation. Am. J. Pharm. Educ. **78**(5), 94 (2014)

28. Sundler, A.J., Pettersson, A., Berglund, M.: Undergraduate nursing students' experiences when examining nursing skills in clinical simulation laboratories with high-fidelity patient simulators: A phenomenological research study. Nurse Educ. Today **35**(12), 1257–1261 (2015)

Workshop on Technology-Enhanced Learning in Digital Creativity Education (TEL4creativity)

Workshop on Technology-Enhanced Learning in Digital Creativity Education: TEL4creativity (TEL4creativity)

Workshop on Technology-Enhanced Learning in Digital Creativity Education: TEL4creativity (TEL4creativity)

The aim of TEL4creativity is to connect people interested in the themes of creativity, digital creativity and technologies giving them the opportunity to compare their methodologies and practice. In particular, we expect researchers, psychologists, educators and technology experts to give their contribution in sharing their expertise in a multidisciplinary context to discuss new tools and propose innovative methodological approaches. In the context of school and teachers education the theme of digital creativity is lively debated as it is promoted at institutional level (consider the indication for European Community) but it is difficult to put it in practice at different levels, ranging from needed infrastructure, to teachers preparation and student involvement. It is therefore a challenge to find the methodologies and identify the tools that can be fit, including MOOC, Serious Games, Tangible User Interface, Robotics, etc.

This workshop invites researchers and practitioners to submit research papers and work in progress investigating methodological and technological aspects related to digital creativity education.

We encourage particularly the submission of articles (both long and short paper) regarding TEL for digital creativity from the point of view of teachers (TEL for digital creativity in curricula, digital creative resources, etc.) and students (digital creation and expression, divergent thinking, computational thinking, etc.) from the point of view of different experts: educators, psychologists, technology experts, game designer and developers, etc.

The topics of the workshop include (but are not limited to):

- Design and implementation of digital creative learning scenarios
- TEL for creativity in formal and informal contexts
- Usability and integration of different TEL tools for digital creativity
- Intelligent systems for adaptation to users
- Teacher and student tools in TEL in digital creativity

- Evaluation methods and techniques for evaluating digital creative learning
- Methodological framework for TEL in digital creativity
- The interplay of TEL and Digital Creativity
- Approaches for learning in Digital Creativity
- Tools that support learning design evaluation
- Assessment of digital creativity in students
- Peer and self-assessment in TEL and digital creativity

Organization

Organizing Committee

Raffaele Di Fuccio	University of Naples Federico II, Italy
Luigia Simona Sica	University of Naples Federico II, Italy
Michela Ponticorvo	University of Naples Federico II, Italy

Program Committee

Michela Ponticorvo	University of Naples Federico II, Italy
Orazio Miglino	University of Naples Federico II, Italy
Raffaele Di Fuccio	Smarted s.r.l., Italy
Luigia Simona Sica	University of Naples Federico II, Italy
Mario Barajas	University of Barcelona, Spain
Frederique Frossard	University of Barcelona, Italy
Anna Trifonova	CreaTIC Academy S.L, Spain

Creating Digital Environments for Interethnic Conflict Management

Elena Dell'Aquila[1]([⊠]), Federica Vallone[2], Maria Clelia Zurlo[1],
and Davide Marocco[2]

[1] Department of Political Science, University of Naples, Federico II,
Via Rondinò 22, 80138 Naples, Italy
{elena.dellaquila, zurlo}@unina.it
[2] Department di Humanistic Sciences, University of Naples, Federico II,
Via Porta di Massa 1, 80133 Naples, Italy
{federica.vallone, davide.marocco}@unina.it

Abstract. Role play simulation games have recently received attention in the training and education fields as a mechanism for providing generative and creative learning. E-learning systems can provide a solid platform upon which role play simulation can be created and implemented to promote both knowledge and competence development. This form of active learning provides a unique tool for training people in different context of applications which may be able to benefit from the availability of open source e-learning tools and overcome lack of access to affordable training and developmental resources. In this paper it will be described the approach and the methodology used to develop a digital single player role play game, named ACCORD, so to provide a new e-learning and flexible tool, to help teachers and educators to autonomously improve and assess their intercultural competencies within the school context.

Keywords: Virtual role-play · Creative learning · Negotiation ·
Interethnic conflict resolution · Simulation

1 Introduction

Within the fields of training, virtual educational role-playing games are receiving recognition for their capacity to provide generative learning. Such trainee-centered learning activities can indeed readily facilitate "learning by doing" as the training scenarios can be developed simulating real life situations, and through their verisimilitude, can enable the transfer of what has been learnt to similar real-life contexts (Ferrara et al. 2016; Pacella et al. 2015). This paper explores the methodological and technological aspects underpinning the training of effective communication and conflict management skills through the use of a Technologically Enhanced Educational Role-Playing Game (EduTechRPG), ACCORD developed within the Erasmus+ ACCORD project (http://accord-project.eu).

Intelligent tutorship, psychological modelling, and feedback mechanisms for ensuring the success of the learning process, represent the fundamental characteristics of this methodological approach for training soft skills as interethnic conflict management.

© Springer Nature Switzerland AG 2020
E. Popescu et al. (Eds.): MIS4TEL 2019, AISC 1008, pp. 81–88, 2020.
https://doi.org/10.1007/978-3-030-23884-1_11

It will be described how the modelisation and operationalisation of theoretical dimensions related to effective conflict management (models of handling interpersonal conflict and assertive communication) can be embedded and implemented in the ACCORD EduTechRPG so to define and design a methodology able to train the user's interethnic negotiation and communication skills in realistic school context scenarios during the interaction with artificial agents (BOT). EdutechRPG represent an innovative form of active learning able to provide a variety of group targets with unique open source e-learning tools for assessment and training of transversal competences to which they can benefit, overcoming the common lack of access to affordable training and developmental resources (Dell'Aquila et al. 2016).

2 The ACCORD Project

The main motivation for the ACCORD project comes from a need emerged at European level, for flexible, accessible, and affordable learning tools, which are capable of helping a large majority of people working in the educational sector to access informal learning pathways for the autonomous improvement and assessment of their social competences in relation to the ever changing challenges posed by the active pursuit of cultural integration in schools and other educational settings, which often passes through difficult stages made of intercultural clashes and conflicts. In this respect, the ACCORD project aims to offer a structured and innovative platform for the provision of pedagogy and assessment methods for helping teachers and educators to enhance and (self-)assess their intercultural competences and negotiation abilities.

The training methodology is based on the production of theoretical training material on the subject through a MOOC (Massive Open Online Course) offering multimedia material and video lecture by various experts on the topics and the development of a series of virtual role-playing game scenarios where teachers can directly experience the dynamics of interethnic conflicts.

ACCORD aims to upscale and customize the ENACT training methodology based on a single-player 3D role-play game intelligence-based tool, developed in a previous EU project (www.enactgame.eu), to train and assess users' negotiation and communication skills in realistic scenarios during the interaction with artificial agents to be employed within enterprises, professional training and sport contexts.

This platform, based on recent psychological modelling through the application of current ICT research such as e-learning, mobility, internet, artificial intelligence (Pacella et al. 2017; Ponticorvo et al. 2016), within the ACCORD project will be tested in five European countries such as Austria, Belgium, Germany, Italy, Spain through the definition, design and implementation of tailored virtual training scenarios based on dynamics between teacher-pupil interaction occurring during interethnic conflict management and resolution. To do so, focus groups with representative of secondary school teachers have been organized across the five EU country so to explore, investigate and identify current issues around these topics. During focus groups we looked at how the active participation of teachers in the process of designing significative learning scenarios may have a great impact on the creation of meaningful learning experiences transferable to the school context.

With this regard, one of the main purposes of ACCORD is to provide teachers with tailored knowledge on interethnic conflict resolution. During focus groups they have contributed to the creation of the content of the virtual learning scenarios simulating interethnic conflicting situations, that is: defining what may be considered as an interethnic conflict and outlining possible scenarios based on their experience; identifying what are the main feelings that may be experienced both by teachers and students involved; setting up what kind of interventions may be considered effective to redeem the proposed conflict; and analyzing what is the most effective communication style to be used to deal with a specific situation, context and student. Therefore, playing the virtual learning scenarios in class has as main objective to boost awareness around effective intercultural communication in order to help teachers to experience how to manage conflict constructively, so to create positive learning environments and facilitate students in developing conflict resolution competences too.

Indeed, negotiation and communication skills are considered fundamental aspects that can help teachers to effectively foster the inclusion of disadvantage learners, including people with a migrant background, while preventing and combating discriminatory practices. Once completed a scenario the user's skill are assessed by an intelligent artificial tutor by using the most advanced methodologies in artificial cognitive systems which offers a profile regarding the negotiation and communication path adopted by a certain teacher during the game.

2.1   The ACCORD Methodology

The design of the ACCORD game scenarios follows the EduTechRPG methodological approach defined as Simulation Technology-drama based system (Dell'Aquila et al. 2016).

These aspects reflect the following two main dimensions: (1) psycho-pedagogical and (2) technological. The psycho-pedagogical dimension specifies the foundations of the learning approach adopted. In the case of the ACCORD game this allows users to experience direct involvement within the learning environments/scenario, where they act through a personal dramatization by using a digital alter-ego, similarly to what occurs in face to face role-play activities; with regard to the technological dimension, ACCORD is based on a technological approach so called SimTech that permits the production of "artificial" micro-worlds based on computer simulated, formal, models about social and psychological phenomena to which users interact within.

In general terms, the core elements of the EduTechRPG methodology resides in the gaming experience, underpinned by the modelling of relevant psychological and pedagogical theories and the use of simulation and artificial intelligence techniques, and the tutoring and assessment of users.

The ACCORD game scenarios can be characterized by the presence of a real or virtual trainer (the latest expression of a computational model), also referred as back stage agent (BSA), as she does not intervene and affect directly the dynamics of the game, though support of its interpretation and wielding. Conversely the actors of the game whether real or artificial represent the on-stage agents (OSA) because it is only through their choices that the game comes to life, and their actions can change the state of the game.

The starting point for designing any educational virtual scenario is the soft skill to be trained or assessed, that in our case is interethnic conflict management.

Given that, the structure for the design and implementation of the game follows the definition of three interrelated architectural elements:

(1) visible layer (core of gameplay); (2) hidden layer (theoretical principles to be transferred); and (3) evaluation layer (users' assessment), where the tutor can be either virtual of real.

The visible and the hidden layers present a general structure that can be found in any other games (i.e. entertainment commercial games). The evaluation layer is instead specifically related to the design and implementation of the game and is therefore a peculiarity of educational/training games.

More precisely visible layer is what the user sees and acts upon during the game, the narrative that introduces the player to the game and contributes to the development of an immersive atmosphere. The visible layer represents the "gamification aspect" of the design and is strictly related to the hidden layer, so that the user is provided with the correct level regarding the content of training and the gaming operations to play with. Visible and hidden layer together, are the funding aspects for the educational objectives of the game and permit the user to have a first-hand experience of the concepts to be learnt. In other words, if the ACCORD game aims at boosting intercultural negotiation competences, the user, regarding the visible layer, should be able to experience negotiation-based activities, and exercise related relevant skills such as effective and assertive communication.

The evaluation layer (complementing the hidden and visible layers) has the role of analysing user's game performances according to the specified training objectives and to provide the user with meaningful information about the learning process and overall performances throughout the duration of the game.

The evaluation layer is based on a combination and integration between the assessment methodology proper for entertainment games (learning analytics and educational data mining to progress through levels), and the traditional role-playing methodology based on the provision of specifically designed tools for tutoring and assessment, consisting in the feedback and debriefing processes.

In the case of ACCORD, in order to produce a reliable educational virtual game in such a domain it is therefore instrumental that the hidden layer takes into account the modalities in which the user can experience the dialogue with another person under precise circumstances. As we will see, in effective communication within a conflicting scenario, besides the actual meaning of the sentences, non-verbal and para-verbal communication play a crucial role. Therefore, it is essential that the visible layer presents the user with the correct level of information (i.e. content of the sentences, body language, facial expressions) and the hidden layer implements a number of rules and dynamics to promote interesting and meaningful experiences for users. For this reason, we will directly involve secondary school teachers in the creation and definition of the interethnic conflict scenarios in classroom.

2.2   The ACCORD Game

The ACCORD game focuses on the simulation of a dialogue between two characters (an avatar controlled by a human player, and a BOT computer-controlled counterpart), in which behavioural characteristics such as the act of speech and some elements of body language play a fundamental role (Marocco et al. 2015). In this regard and according to the methodological dimensions delineated previously, ACCORD can be seen as an implementation of Drama-based games. Moreover, it is designed to be played by a single user over the Internet, and the interaction user-3D artificial agent (BOT) is simulated on the computer according to well-defined psychological theories. This aspect characterises the platform as SimTech.

More specifically the ACCORD hidden layer and its simulation role play game scenarios are based on models of handling interpersonal conflict (Rahim and Bonoma 1979) and assertive communication in effective interactions (Dryden and Constantinou 2004).

As described following, the underlying psychological models of communication and negotiation are based on a consolidated negotiation model inspired to the five styles of handling conflict developed by Rahim (Rahim and Bonoma 1979) that are differentiated on two basic dimensions: concern for self and concern for others. The first dimension explains the degree (high or low) to which a person attempts to satisfy his or her own concern. The second dimension explains the degree (high or low) to which a person attempts to satisfy the concern of others. The combination of the two dimensions results in five styles of handling interpersonal conflict:

- Integrating style (high concern for self and for others), that involves collaboration between teacher and student that are willing to reach a mutual and acceptable solution through exchange of information, examination and exploration of differences.
- Obliging style (low concern for self and high concern for others), also known as accommodating style, is associated with attempting to play down the differences and emphasizing commonalities to satisfy the concern of the other party (the student). However, this style can be useful when preserving a relationship with the student might be the objective of the teacher, yet as a strategy when he is willing to give up something with the hope of getting some benefits in the future.
- Dominating style (high concern for self and low concern for others) is associated with forcing behaviour to win one's position. This style may be an appropriate choice when an immediate urgent action is needed, or when an unfavourable decision taken by student involved in a conflict can be harmful to this party itself or the classmates.
- Avoiding style (low concern for self and others) is associated with withdrawal and may take the form of postponing an issue until a better time, or simply retreating from a threatening situation. A teacher using an avoiding style can be hardly be considered to constitute an appropriate style of handling intercultural conflicts.
- Compromising style (intermediate in concern for self and others) involves give-and-take whereby both parties give up something to reach a mutually acceptable decision.

The ACCORD platform has been designed to bring Rahim's model into a 3D virtual environment. The game will be organized in different interethnic conflict scenarios which content has been identified and discussed with teachers during focus groups, where the user plays a role of the teacher and negotiates with various virtual agents representing students in interethnic conflicting situations.

The digital characters representing the user and the artificial agent are implemented as 3D avatars. These can perform a range of basic expressions using verbal cues (vocal tone, shape of the speech bubble and structure of the sentence) and non-verbal indicators (facial expression, eye contact, body posture and gestures). The user is introduced to the game with a scene explaining the situation and the reason of the conflict that conduct the player to the stage of the virtual agent-user interaction (Fig. 1).

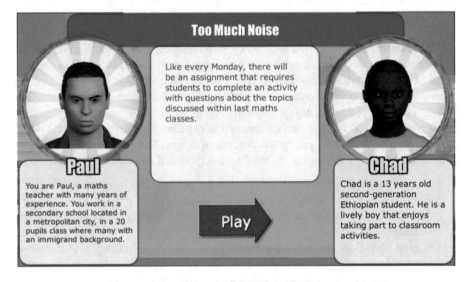

Fig. 1. An example of ACCORD conflicting scenario

Each exchange between the two is organized in a five-state scene, which include one turn of speech for each party. In each state, the user can choose one among five possible sentences (one for each of Rahim's styles of handling conflicts) that are complemented with gesture and facial expression as expression of the non-verbal indicators as identified by the assertive model of communication for the assertive, passive and aggressive behavioural styles. Both verbal and non-verbal indicators in the communication process between the user and the artificial agent have been chosen for their relevance in the behavioural description of the three different communication styles, and they can be seen as indicators of specific behavioural traits that can be objectively observed and measured.

Once a scenario has been completed the teacher is given: (1) a debriefing regarding the main path of negotiation styles adopted during the games); and (2) a profile based on the Rahim model that is related to the specific situations she played and the efficacy of the communication style acted throughout the exchange with the student (Fig. 2).

The user is also provided with an outline regarding the history of all the choices made and guided through the understanding of different aspects of the negotiation.

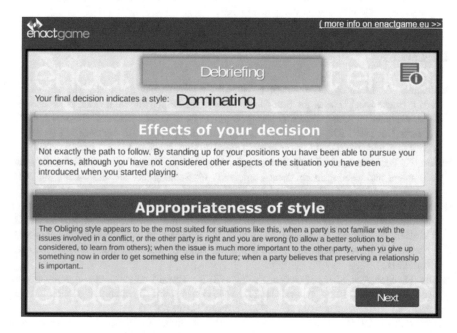

Fig. 2. Example of users' negotiation path and users' profile

3 Conclusions and Future Work

The ACCORD project methodology aims to train secondary school teachers across the five different project countries so to prepare them to take an active stand against discrimination and racism, and to deal with diversity and handle intercultural conflicts. This will be achieved through teachers' direct involvement in the design of the content of the ACCORD game scenarios followed by hands-on experience on the concepts and dynamics of negotiations simulated into the such scenarios.

Indeed, game design and experience will be accompanied with a strong theoretical background provided by the development of a MOOC course supporting teacher with in-depth multimedia and video lecture material available online to develop competences related to intercultural literacy, inclusive pedagogical approaches and conflict management, creation of positive learning environments featured by intercultural coexistence and positive approaches towards conflicts.

The design of game-based learning scenarios passing through the direct participation and contributions of teachers may provide situated learning opportunities both for teachers and educational practitioners engaged to deal with a school system becoming increasingly more linguistically and culturally diverse, due to the rapid growth of immigrant flows which has occurred over the last decade in Europe.

Acknowledgments. This research has been supported by the EACEA, project ACCORD (580362-EPP-1-2016-1-IT-EPPKA3-IPI-SCO-IN) from the Erasmus+. This publication reflects the views only of the author, and the Commission cannot be held responsible for any use which may be made of the information contained therein.

References

Dell'Aquila, E., Marocco, D., Ponticorvo, M., Di Ferdinando, A., Schembri, M., Miglino, O. (eds.): Educational Games for Soft-Skills Training in Digital Environments: New Perspectives. Advances in Game-Based Learning. Springer (2016). E-book ISBN 978-3-319-06311-9. Hardcover ISBN 978-3-319-06310-2. https://doi.org/10.1007/978-3-319-06311-9

Dryden, W., Constantinou, D.: Assertiveness Step by Step. Sheldon Press, London (2004)

Ferrara, F., Ponticorvo, M., Di Ferdinando, A., Miglino, O.: Tangible interfaces for cognitive assessment and training in children: LogicART. In: Smart Education and e-Learning 2016, pp. 329–338. Springer, Cham (2016)

Marocco, D., Pacella, D., Dell'Aquila, E., Di Ferdinando, A.: Grounding serious game design on scientific findings: the case of ENACT on soft skills training and assessment. In: Conole, G., Klobučar, T., Rensing, C., Konert, J., Lavoué, É. (eds.) Design for Teaching and Learning in a Networked World. Lecture Notes in Computer Science, vol. 9307, pp. 441–446. Springer, Cham (2015)

Pacella, D., Dell'Aquila, E., Marocco, D., Furnell, S.: Toward an automatic classification of negotiation styles using natural language processing. In: International Conference on Intelligent Virtual Agents, pp. 339–342. Springer, Cham (2017)

Ponticorvo, M., Di Ferdinando, A., Marocco, D., Miglino, O.: Bio-inspired computational algorithms in educational and serious games: some examples. In: European Conference on Technology Enhanced Learning, pp. 636–639. Springer, Cham (2016)

Rahim, M.A., Bonoma, T.V.: Managing organizational conflict: a model for diagnosis and intervention. Psychol. Rep. **44**, 1323–1344 (1979)

The Importance of Spatial Abilities in Creativity and Their Assessment Through Tangible Interfaces

Antonio Cerrato[1]([✉]), Giovanni Siano[2], Antonio De Marco[2], and Carlo Ricci[3]

[1] Department of Humanistic Studies, University of Naples "Federico II", Naples, Italy
antonio.cerrato@unina.it
[2] Quing System, Agropoli, Italy
{giovanni.siano,antonio.demarco}@quing.it
[3] Walden Institute, Rome, Italy
presidenza@istituto-walden.it

Abstract. Spatial abilities play a key and unique role in structuring many important psychological phenomena including creativity, a quality that in the past was considered for few gifted people.

In this paper we examine the important relationship between creative thinking and spatial cognition, and how particular spatial skills can impact on some aspect of the creative process. Moreover, after a brief mention on the neural basis of creativity we present a prototype, supporting tangible interfaces, with which is possible to create tasks that assess and train those spatial abilities linked with creative thinking.

Keywords: Spatial abilities · Creativity · Assessment · Tangible user interfaces

1 Introduction

1.1 The History of the *Creative* Brain

In the past, creativity has been considered as a rare and mysterious trait, but nowadays it is considered like an heritage of all individuals. We moved from considering it an innate gift to the discovery of a possible acquisition of this characteristic, and this quality, once considered the exclusive peculiarity of a select few "geniuses", has now become a general humanitarian patrimony.

For a long time creativity has been assimilated to intelligence, but thanks to Torrance [1], were developed tests of measuring the IQ and the most precise creative talent, which led to the conclusion that creative people have an average IQ. Beyond a certain limit the IQ is not fundamental in triggering creativity, it is necessary but not sufficient.

Nowadays, most scholars agree in defining creativity as a production of the new ideas that are transformed into something innovative, impactful from a social point of view, agreeing with the standard definition of creativity, which

© Springer Nature Switzerland AG 2020
E. Popescu et al. (Eds.): MIS4TEL 2019, AISC 1008, pp. 89–95, 2020.
https://doi.org/10.1007/978-3-030-23884-1_12

was originally proposed by Stein in 1953 [2], according to which *"Creativity requires both originality and utility"* and regards the production of something new and useful.

From a psychological point of view, the definition of creativity is referred to Guilford [3], who recognized the idea that the aspects that distinguish creative thinking are:

- **fluidity**, that is the ability to produce abundant ideas, without reference to the their adequacy for the resolution of the problem;
- **flexibility**, that is the ability to change and adapt the ideation process, then to move from one succession of ideas to another, from one scheme to another;
- **originality**, which consists in the ability to find unique, particular and unusual answers;
- **elaboration**, that consists in focussing on each solution/idea and developing it with details;
- **sensitivity to problems**, e.g. selecting ideas and organizing them in new ways, understanding what is wrong and what can be perfected in everyday objects.

Moreover, authors have since extended the thesis that a creative act is not a single event, but a process of interaction between cognitive and affective elements. In this perspective, the creative act has two phases, one generative and one explorative/evaluative [4]. During the generative process, the creative mind imagines a series of new mental models as potential solutions to a problem, in the exploratory phase, the different options are evaluated and then the best one is selected.

The first scholars interested in creativity, like Guilford, characterized two phases of the creative thoughts, the divergent thinking and the convergent thinking [3], defining the first as the ability to produce a wide range of associations or to arrive at many solutions in reference to a problem, it therefore goes beyond what is contained in the starting situation, overcomes the closure of the problem data, explores various directions and produces something new. On the contrary, convergent thinking refers to the ability to quickly concentrate on the best solution to a difficulty, therefore it remains circumscribed within the patterns of the problem.

The idea of the existence of two phases of the creative process is consistent with the results of cognitive research that indicates the existence of two different ways of thinking, one associative and one analytical [5,6]. In associative mode, the thinking process is intuitive, differently, in analytical mode, thoughts are focused on cause-effect relationships.

Ultimately, creativity seems not to be a unique property, but is the result of the mixed process between deduction and intuition, between reason and imagination, between emotion and reflection, between divergent and convergent thinking.

1.2 Neural Basis of Creativity

Although a series of activities have been carried out in the field of cognitive neuroscience, recent research has shown that there is no coherent and unified framework for the neuroanatomical bases of creativity. Martindale [7] has indagated the role of hemispheric asymmetry in creativity, due to fact that creative process was considered a function of the right hemisphere, in contrast to the rational process managed by the left hemisphere. This theory, fascinating for its linearity, appears today simplistic, compared to the complexity of the brain, for example the investigation on divergent thinking using the electroencephalography (EEG) have not confirmed a specific right lateralization of creativity [8].

Functional neuroanatomical studies, performed using functional magnetic resonance imaging (fMRI) or positron emission tomography (PET) have reported greater involvement of the prefrontal cortex (PFC) as a critical neural substrate for divergent thinking [8,9], an evidence confirmed also from other researches reporting a constant activation of the PFC in creative thinking tasks [10,11].

Also other cerebral structures, both cortical and subcortical, seem to play a role in divergent thinking such as the visual areas [12], the striatum [13], the anterior cingulate gyrus [10,12], the cerebellum [14] and the corpus callosum [15]. However, these results are dispersive and apparently are not supported by the overwhelming majority of studies.

Nevertheless there is no unanimous agreement among researchers on what are the precise brain areas involved in the creative process, the involvement of the PFC remains clear. Moreover, there is not a specific set of brain regions involved in this process, but rather a widespread activation of different brain areas [16]. In addition, although further studies are necessary to identify specific brain areas for creative processes, some cerebral structures described above are also involved in spatial or visuospatial tasks; in fact, some researches show a connection between spatial abilities and creative thinking and it will be discussed in the next section.

2   Spatial Cognition and Creativity

Commonly, people have always sustained the existence of a connection between creativity and spatial cognition, starting from the hemispheric asymmetry theory of the brain sustaining that both creative thinking and spatial skills would be processed by the right hemisphere [17]. Moreover, spatial cognition is been also linked with general intelligence [18], and it is considered particularly crucial for the success in specific learning field such as geometry, mathematics and other STEM disciplines [19]. However, as concerns spatial abilities, it has to be said that they rely on many different factors. First of all, identifying and locating objects in the external world through the vision is a fundamental for spatial cognition. In particular, vision refers to the perception of shapes, colors, distances, patterns, three-dimensional geometric forms, line drawings and so on [20].

Starting from the basic perception of spatial relationships, other spatial abilities (e.g. visualization or mental imagery [21]) can be possessed at different levels of expertise and some authors investigated their link with creativity. Kozhevnikov et al. [22] have been involved in the study of the relationship between visualization and creativity, sustaining the importance of visualization abilities for the creative dimension, (especially in applied fields, such as visual art/science education) and revealing that art students were better than non-artists in noticing pattern alterations in pictures. This finding is inline with a previous research of Kozbelt [23] showing that art students performed significantly better in mental imagery tasks. Another research [24] has showed the correlation between mental imagery in divergent thinking, that is one of the characteristics of creative thoughts. Considering the importance of spatial cognition in creativity, recognized also by scientific researches, we propose a prototype to train and to assess spatial skills in order to deepen the knowledge and the relationship with the creative process.

2.1 ETAN: A Technology Enhanced Platform to Train and Assess Spatial Abilities

Here we present ETAN, a platform that supports the use of tangible user interfaces (TUIs, physical manipulable object technologically enhanced) to assess and train spatial abilities; it represents a second version of a tool designed for the evaluation of spatial cognition [25]. In particular, we ideate this prototype to investigate the spatial behaviors of people (children and adults) in relation to their proximal/peripersonal space, that is commonly defined as the space immediately surrounding our bodies [26]. Objects within peripersonal space can be reached by hands and manipulated; objects located beyond this space (in what is often termed 'extrapersonal space') cannot normally be touched without moving toward them, or else their movement toward us.

The tangible parts of ETAN are represented by objects (small disk in our first version of the prototype) that are possible to detected by a camera connected to a PC. These objects are detectable, in a delimited space, thanks to a particular kind of tags, popular in augmented reality technology [27] named ArUco Markers [28], that can be traced by a specific software developed with an artificial vision module; in this manner is possible to recreate on the computer screen the physical disposition of objects and store the data about each session (Fig. 1).

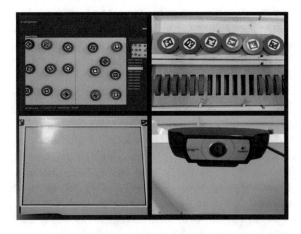

Fig. 1. The prototype ETAN

Thanks to the use of ArUco Markers, each physical object can be artificially enhanced, and this allows the creation of different activities related to spatial abilities. For example, it is possible to develop tasks in which people are asked to explore, through the positioning of objects, different points of the peripersonal space, or to compare simple spatial dichotomy concepts such as 'Top vs Bottom', 'Left vs Right' and so on. Other activities could regard the teaching of shapes and colors (using objects with different forms and different paints), or even promote creative thinking by asking people to recreate complex figures using the simple shapes of the objects made available to them (for example, the construction of a rectangle starting from two squares, a rhombus with triangles etc.).

We are still exploring the possible use of our platform but we aim to utilize it in assessment and learning contexts, considering it an useful tool to test/train spatial behaviors and divergent thinking of people.

3   Conclusions

In this work we discussed about the existing relationship between spatial cognition and creativity and we proposed a tool to assess spatial abilities; our prototype ETAN (still under development), allows the evaluation of spatial skills through the interaction of users with tangible user interfaces.

Visuospatial competence serves as support of the creative process and innovative ideation, and their role in creativity is supported by various scientists [20]. For example, the proficiency of people in spatial tasks involving visualizing or mental imagery can give many different perspectives of the same problem, stimulating multiple creative ideas. Some doubts remain whether the relationship between spatial ability and creativity reflects inborn predisposition or is a result obtained by the experience. Different from past theories, creativity is considered no more a benefit for few talented people but a characteristic within the reach of everyone.

Therefore, an interesting direction for future research is to investigate how work/discipline preferences are linked to the possess of creative skills and spatial abilities. Finally, different approaches should also be pursued more systematically in order to better evaluate the role of different variables in establishing the complex connection between creativity and spatial ability, and to deepen the knowledge about which brain areas are involved in the creative thinking.

References

1. Torrance, E.P.: The nature of creativity as manifest in its testing. In: The Nature of Creativity: Contemporary Psychological Perspectives, vol. 43 (1988)
2. Stein, M.I.: Creativity and culture. J. Psychol. **36**, 311–322 (1953)
3. Guilford, J.P.: Creativity. Am. Psychol. **5**, 444–454 (1950)
4. Finke, R.A., Ward, T.B., Smith, S.M.: Creative cognition: theory, research, and applications (1992)
5. Neisser, U.: The multiplicity of thought. Br. J. Psychol. **54**, 1–14 (1963)

6. Sloman, S.A.: The empirical case for two systems of reasoning. Psychol. Bull. **119**, 3 (1996)
7. Martindale, C.: Handbook of Creativity, pp. 211–232. Springer (1989)
8. Dietrich, A., Kanso, R.: A review of EEG, ERP, and neuroimaging studies of creativity and insight. Psychol. Bull. **136**, 822 (2010)
9. Folley, B.S., Park, S.: Verbal creativity and schizotypal personality in relation to prefrontal hemispheric laterality: a behavioral and near-infrared optical imaging study. Schizophr. Res. **80**, 271–282 (2005)
10. Fink, A., et al.: The creative brain: Investigation of brain activity during creative problem solving by means of EEG and fMRI. Hum. Brain Mapp. **30**, 734–748 (2009)
11. Gibson, C., Folley, B.S., Park, S.: Enhanced divergent thinking and creativity in musicians: a behavioral and near-infrared spectroscopy study. Brain Cogn. **69**, 162–169 (2009)
12. Howard-Jones, P.A., Blakemore, S.-J., Samuel, E.A., Summers, I.R., Claxton, G.: Semantic divergence and creative story generation: an fMRI investigation. Cogn. Brain Res. **25**, 240–250 (2005)
13. Blom, O., et al.: Regional dopamine D2 receptor density and individual differences in psychometric creativity. Poster Session Presented at the 14th Annual Meeting for the Organization for Human Brain Mapping, Melbourne, Australia (2008)
14. Chávez-Eakle, R.A., Graff-Guerrero, A., García-Reyna, J.-C., Vaugier, V., Cruz-Fuentes, C.: Cerebral blood flow associated with creative performance: a comparative study. Neuroimage **38**, 519–528 (2007)
15. Moore, D.W., et al.: Hemispheric connectivity and the visual-spatial divergent-thinking component of creativity. Brain Cogn. **70**, 267–272 (2009)
16. Dietrich, A.: Introduction to Consciousness. Macmillan International Higher Education, Basingstoke (2007)
17. Rubenzer, R.: The role of the right hemisphere in learning & creativity implications for enhancing problem solving ability. Gift. Child Q. **23**, 78–100 (1979)
18. Gardner, H.: Frames of Mind: The Theory of Multiple Intelligences. Hachette, London (2011)
19. Lubinski, D.: Spatial ability and STEM: a sleeping giant for talent identification and development. Pers. Individ. Differ. **49**, 344–351 (2010)
20. Palmiero, M., Srinivasan, N.: Creativity and spatial ability: a critical evaluation. In: Cognition, Experience and Creativity, pp. 189–214, April 2015. ISBN 978-81-250-5731-4
21. Finke, R.A.: Creative Imagery: Discoveries and Inventions in Visualization. Psychology press, New York (2014)
22. Kozhevnikov, M., Kozhevnikov, M., Yu, C.J., Blazhenkova, O.: Creativity, visualization abilities, and visual cognitive style. Br. J. Educ. Psychol. **83**, 196–209 (2013)
23. Kozbelt, A.: Artists as experts in visual cognition. Vis. Cogn. **8**, 705–723 (2001)
24. LeBoutillier, N., Marks, D.F.: Mental imagery and creativity: a metaanalytic review study. Br. J. Psychol. **94**, 29–44 (2003)
25. Cerrato, A., Ponticorvo, M.: Enhancing neuropsychological testing with gamification and tangible interfaces: The Baking Tray Task. In: International Work-Conference on the Interplay Between Natural and Artificial Computation, pp. 147–156 (2017)
26. Rizzolatti, G., Fadiga, L., Fogassi, L., Gallese, V.: The space around us. Science **277**, 190–191 (1997)

27. Cerrato, A., Siano, G., De Marco, A.: Augmented reality: from education and training applications to assessment procedures. Qwerty-Open Interdiscip. J. Technol. Cult. Educ. **13** (2018)
28. Garrido-Jurado, S., Muñoz-Salinas, R., Madrid-Cuevas, F.J., Marín- Jiméenez, M.J.: Automatic generation and detection of highly reliable fiducial markers under occlusion. Pattern Recogn. **47**, 2280–2292 (2014)

The DoCENT Game: An Immersive Role-Playing Game for the Enhancement of Digital-Creativity

Raffaele Di Fuccio[1]([⊠]), Fabrizio Ferrara[2], and Andrea Di Ferdinando[1]

[1] Smarted srl, Via Riviera di Chiaia 256, 80121 Naples, Italy
raffaele.difuccio@gmail.com
[2] Università di Napoli Federico II, Via Porta di massa 1, 80133 Naples, Italy

Abstract. The DoCENT project (Digital Creativity ENhanced in Teacher education) co-funded by the Erasmus+ programme of the European Union aims to enhance digital creativity in Initial Training Education (ITE) contexts. The paper shows one of the outputs of the project, namely the Serious Game developed as a role-playing game where the game proposes a natural and realistic interaction with the class and poses the user (i.e. the teacher or a generic tutor) in front of problematic situations for an immersion in a real classroom experience. In the paper the authors describe the prototypal DoCENT game and discuss about the model that includes five steps. In particular the focus in on the co-creation process in order to design the learning scenarios with the teachers and the real users of the Serious Game. In addition, the paper shows the methodology of the learning scenarios, the interaction of the players and the feedbacks derived from an adaptive tutoring system.

Keywords: Serious game · Digital creativity · Role-playing game ·
Adaptive tutoring system · Learning scenarios · Initial Training Education

1 Introduction

1.1 The Digital Creativity

In a digital world, the digital skills are basic in the workplaces and in each business context in order to increase the productivity and the management of new business opportunities. Every day, people have to face with digital technologies both if they are digital native, both if they have a low degree of digital skills. The digital competences have gained more importance in our society and it confirmed also by the huge commitment of the European Commission that has released a competence framework named DigiComp[1] that now is at the 2.1 release [3].

At the same time, the creative skills are being even more fundamental for the challenge of the XXI Century [2, 4]. The digital technologies propose new schemes and paradigms that need new solutions to be faced. It is necessary a creative approach for a

---

[1] https://ec.europa.eu/jrc/en/publication/eur-scientific-and-technical-research-reports/digcomp-21-digital-competence-framework-citizens-eight-proficiency-levels-and-examples-use.

E. Popescu et al. (Eds.): MIS4TEL 2019, AISC 1008, pp. 96–102, 2020.
https://doi.org/10.1007/978-3-030-23884-1_13

proper use of the digital technologies. In order to affirm this, also the DigiComp framework define a specific item for the application of the creativity with the digital technologies with the aim "to use digital tools and technologies to create knowledge and to innovate processes and products" and "to engage individually and collectively in cognitive processing to understand and resolve conceptual problems and problem situations in digital environments". Also, innovative workforce requires both the ability to work with technologies and to adapt, generate new ideas, products and practices [16]. The integration between the creative skills in the application of digital technologies has raising its importance and a crucial educational objective [8].

This increasing interest of EU sets new policies and brings this conjunction between creativity and digital technologies in the educational settings but it has a limited effect in the digital economies [16]. The real limit of this framework is related to the low preparation of the teachers. Workers in the education sector are 15% points less likely to have good ICT and problem-solving skills than those working in the professional, scientific and technical activity sectors, which includes scientific research and development and legal and accounting activities [14]. With this background the adaptation of pedagogical strategies for fostering creativity are strongly decreased.

At the same time, teacher educators are key players for ensuring the quality of teaching professions and the support of educational innovation[2] and this is applicable also to the new generations of teachers and then in the initial teacher education (ITE). In particular, in the ITE curricula the application of creativity in conjunction with digital technologies are not deeply explored.

The project DoCENT (Digital Creativity ENhanced in Teacher education) aims to face this challenge, focusing on initial teacher education (ITE), namely the new generation of teachers, aiming to enhance digital creativity in ITE contexts. In this view, the definition of the digital creativity [1] of the DoCENT project is the use of digital technologies to develop processes that address to creativity. Applied in the education field, digital creative teaching consists of applying digital pedagogies to develop processes that are particular to creativity, i.e. promoting learner-centred methodologies, allowing for self-learning, helping to make connections, boosting exploration and discovery, providing a safe environment that encourages risk-taking behaviours and encouraging collaboration.

The project, funded by the European Commission in the Erasmus+ programme that is ongoing, aims to develop, implement, validate and disseminate an innovative model to guide teacher educators in applying digital creative teaching practices.

The project outputs will be implemented in 3 higher educational systems (i.e. Italy, Greece and Spain), learning cultures and socio-economic situations.

In order to objective will deliver a serious game in a form of role-playing game (RPG).

---

[2] European Commission. (2013). Supporting Teacher Educators for better learning outcomes. http://ec.europa.eu/education/policy/school/doc/support-teacher-educators_en.pdf.

2   Methodology

2.1   The DoCENT Model and the Co-creation Process

Co-creation has a lot of definitions, because is applied in different fields. For example, Prahalad and Ramaswamy [15] propose the co-creation by acknowledging the changing roles in the theatre of the market: the customers and suppliers implicitly work together in order to set a product costs, based on the typical system between supply-demand.

The co-creation is applied between enterprises and institutions or when there is the relationship between therapist and patient during the psychological meetings. Other examples are the joint venture between companies or some process inside the family. Finally, a lot of creativity studies focus on the generation of new collaborative proposal starting from shared co-creation processes or in studies and collaborative designs that involve different agencies, or during jointing projects (e.g. European Funding projects).

In addition, the co-creation represents a strong paradigm for the collaborative interaction between teachers and students in the education, that is the field where we apply the co-creation model [7]. However, one of the most accepted definition of co-creation is that is represent a joint, collaborative, concurrent, peer-like process of producing new value, both materially and symbolically [11].

The DoCENT project defined a methodology that covers 5 stages and in particular these are the following:

1. Task identification through face-to-face workshops;
2. Preparation MOOC online learning modules with digital creativity technologies;
3. Co-design of learning scenarios;
4. Application of the co-created learning scenarios in ITE;
5. Response validation.

The model is drawn in the Fig. 1 below.

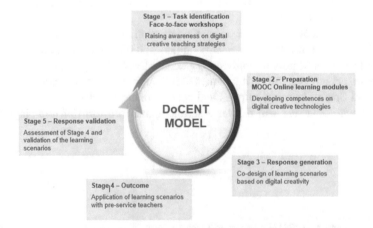

Fig. 1. The five steps of the DoCENT model

In particular the third stage focuses on the co-creation of the learning scenarios in order to process these scenarios in the ITE courses. The pre-service teachers and in-service teachers act a relevant role in the definition of the learning scenarios, providing new solutions, feedbacks and improvements.

This is particularly relevant, because the collaboration among teachers appeared as a key-factor to creativity. Sharing opinions among the teachers enhanced the processes of generating and testing ideas [9].

During the events of the stages 3 [10], the teacher educators are introduced to the concept of digital creativity and to the competence framework defined by the Consortium. In those events the teachers have the chance to meet company stakeholders and learn about digital creativity. Through hands on sessions, they are able to try out digital applications (e.g. GBL, robotics, manipulative technologies) and analyse their educational affordances.

In these co-creation meetings, they had a double role. On one hand, as mediators, they shaped the game dynamics, profiling its mechanisms, and facilitating the production of ideas by providing schemes to design the different game elements. On the other hand, they acted as constraints, since scenarios were conditioned by the characteristics of the software.

3   Technical Description of the Prototype

3.1   The DoCENT Game

The DoCENT Serious Game (DSG)[3] is the output of the project and represent a role-play game.

Role-play encourages new ways of thinking and interacting with things and people of our personal environment. This technique applies different dramatic instruments derived from sociodrama and psychodrama, such as replaying a scene or a part of a scene, role reversal, making asides, mirror and double [5]. These enactment tools facilitate learners to explore their emotions, concepts and thoughts from a detached perspective, and the development of metacognition capacity [17].

In this field, we present the DoCENT Serious game that is planning to be embedded in the MOOC, as a learning tool.

The DSG is browser game for role-play simulations and it is adapted and further improved based on the results achieved from an EU project named ENACT[4] [13] which developed a serious game aimed at training and assessing negotiation soft-skills.

The game proposes an interaction that aims to provide a realistic experience of the organization and management of a real classroom related to digital creative competences. Teacher educators are able to learn how to manage classroom and interact with students following creative pedagogies, i.e. promoting learner-centered methodologies, allowing for self-learning, helping to make connections, boosting exploration and

---

[3] www.docent-project.eu.

[4] www.enact-game.eu.

discovery, providing a safe environment that encourages risk-taking behaviors, encouraging collaboration.

Users can autonomously execute every activity, so that they can experience role-play simulation without the need of a human tutor and immersivity within the game is achieved by a realistic setting. Users are able to learn in a safe environment, anticipating and preventing problems which may appear in the classroom. The virtual tutor allows for observing and analysing the process. At the end of each session, the software provides a recording of users' interaction with the class and reporting comments on the session.

The game is a Window executable file and is in the prototype phase, where one learning scenario is implemented. The structure of the learning scenario is showed in the following Fig. 2.

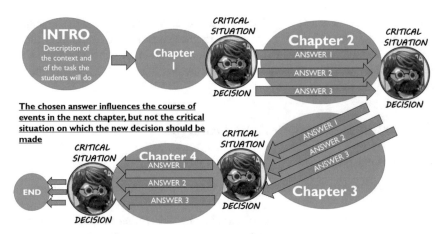

Fig. 2. The structure of the DoCENT learning scenarios for the Serious Game. Each chapter present a critical situation where the player should apply for an answer and proceed in the scenario

The Learning Creativity Scenario is drafted in order to present to the user an activity in a virtual environment for the improvement of the digital creativity competence. The scenario is designed to: (i) propose a natural and realistic interaction with the class, (ii) pose the user (i.e. the teacher or a generic tutor) in front of problematic situations for an immersion in a real classroom experience, (iii) understand the reactions of learners. In each chapter, that represents a classroom situation (i.e. the pupils try to use Sketch but are not able to perform the task) where the teacher should intervene. In each critical situation, the player that has the role of teacher selects an answer in order to face the next chapter.

All the interactions are tracked and an Intelligent and Adaptive Tutoring System (ATS) produces a game interaction that can affect on the reply of the user (i.e. to gain a result there are more than one way to pursuit). ATS will reproduce teachers' function, operating activity selection; adapting interaction; building individualized report on learner interaction. ATS will be based on Learning Analytics modules and Educational Data Mining tools (Fig. 3).

Fig. 3. The image shows two screenshots of the game. On the left a situation during the game, the character speaks with speech balloons, on the right a screenshot of the debriefing where the player obtains a feedback on his/her interaction

4   Conclusion and Future Directions

The DoCENT Serious Game is in a prototypal stage. The paper aims to propose the model of the learning scenarios and structure of the game. The game applies the creativity on digital environment, based on learning scenarios co-created with the real users, in this case teachers. This design model allows a way to implement elements and issues related to routines of the teachers in the classrooms.

The next steps for the implementation of the game will be to implement new learning scenarios that involves other aspects of the digital creativity (the application of tangible user interfaces [6] and the creation of digital games [12]). In addition, the game will be developed as the Android App for smartphone and the browser game.

Acknowledgment. The DoCENT project (Digital Creativity ENhanced in Teacher education) is co-funded by the Erasmus+ programme of the European Union, in the call Key Activity 2 – Strategic Partnership and runs between October 2017 and September 2019.

References

1. Barajas, M., Frossard, F., Trifonova, A.: Strategies for digital creative pedagogies in today's education. In: Active Learning. IntechOpen (2018)
2. Beghetto, R.A.: Creativity in the classroom. In: Kaufman, J.C., Sternberg, R.J. (eds.) The Cambridge Handbook of Creativity, pp. 447–463. Cambridge University Press. Cambridge (2010)
3. Carretero, S., Vuorikari, R., Punie, Y.: DigComp 2.1: the digital competence framework for citizens with eight proficiency levels and examples of use (No. JRC106281). Joint Research Centre (Seville site) (2017)
4. Craft, A.: Creativity and Education Futures: Learning in a Digital Age. Trentham Books, England (2011)
5. Dell'Aquila, E., Marocco, D., Ponticorvo, M., Di Ferdinando, A., Schembri, M., Miglino, O.: Educational Games for Soft-Skills Training in Digital Environments: New Perspectives. Springer, Heidelberg (2016)

6. Di Fuccio, R., Siano, G., De Marco, A.: The activity board 1.0: RFID-NFC WI-FI multitags desktop reader for education and rehabilitation applications. In: World Conference on Information Systems and Technologies, pp. 677–689. Springer, Cham (2017)
7. Dodero, G., Gennari, R., Melonio, A., Torello, S.: Gamified co-design with cooperative learning. In: CHI 2014 Extended Abstracts on Human Factors in Computing Systems, pp. 707–718. ACM, April 2014
8. Ferrari, A., Cachia, R., Punie, Y.: 23. ICT as a driver for creative learning and innovative teaching. In: Measuring Creativity, p. 345 (2009)
9. Frossard, F., Barajas, M., Trifonova, A.: A learner-centred game-design approach: impacts on teachers' creativity. Digit. Educ. Rev. **21**, 13–22 (2012)
10. Frossard, F., Trifonova, A., Barajas, M.: Teachers designing learning games: impact on creativity. In: Video Games and Creativity, pp. 159–183 (2015)
11. Galvagno, M., Dalli, D.: Theory of value co-creation: a systematic literature review. Manag. Serv. Qual. **24**(6), 643–683 (2014)
12. Grizioti, M., Kynigos, C.: Game modding for computational thinking: an integrated design approach. In: Proceedings of the 17th ACM Conference on Interaction Design and Children, pp. 687–692. ACM (2018)
13. Marocco, D., Pacella, D., Dell'Aquila, E., Di Ferdinando, A.: Grounding serious game design on scientific findings: the case of ENACT on soft skills training and assessment. In: Design for Teaching and Learning in a Networked World, pp. 441–446. Springer, Cham (2015)
14. OECD: Innovativing Education and Educating for Innovation. http://www.oecd.org/education/ceri/GEIS2016-Background-document.pdf. Accessed 11 Feb 2019
15. Prahalad, C.K., Ramaswamy, V.: Co-opting customer competence. Harv. Bus. Rev. **78**(1), 79–90 (2000)
16. Sefton-Green, J., Brown, L.: Mapping learner progression into digital creativity (2014)
17. Weinert, F.E., Kluwe, R.H.: Metacognition, Motivation, and Understanding. Erlbau, Hillsdale (1987)

Enhancing Digital Creativity in Education: The Docent Project Approach

Luigia Simona Sica[1(✉)], Michela Ponticorvo[1], and Orazio Miglino[1,2]

[1] University of Naples Federico II, Via Porta di Massa 1, 80133 Naples, Italy
lusisica@unina.it
[2] ISTC-CNR, via San Martino della Battaglia, 44, 00185 Rome, Italy

Abstract. In this paper, we describe an approach to enhance digital creativity in education context, using specific methodologies and tools.

Creativity is a key competence in today work market, but in education context, convergent thinking is what is more rewarded. For this reason there is a gap between these dimensions (school and work) about creativity that can be covered adopting an educational approach promoting creativity, mainly in digital contexts. This approach is delineated in detail describing an on-going European project, Docent, an example of the route that can be followed to promote digital creativity in school context both in the educational process and outcomes.

Keywords: Digital creativity · Education · Teacher education

1 Introduction

Today's work market depends more and more on two aspects: creativity and digital competences [21]. Indeed creativity is considered a crucial element to face and manage this century changes and challenges [7] and new technologies are so embedded in every aspects of our daily life that is not possible to neglect them.

Marrying creativity with new technologies, we obtain what can be called digital creativity [26], as shown in Fig. 1.

These three elements and their connections represent what must be promoted and is receiving a notable attention in work context, but this is not what happens in education context where convergent thinking rather than creativity is rewarded with high grades. In the next sections we will focus on these core elements.

2 Creativity and Digital Creativity

Creativity is a term that indicates the cognitive ability to create and invent something new and valuable. A general definition describes creativity as the ability to generate ideas, insights and solutions that possess the feature to be original and flexible [2, 24] or original and effective [19]. Thus, with 'creativity' it is possible to refer to different dimensions about which there is not a shared consensus. Anyway, we can identify four relevant dimensions, Rhodes four Ps model [18] that highlights different sides of the complex concept:

© Springer Nature Switzerland AG 2020
E. Popescu et al. (Eds.): MIS4TEL 2019, AISC 1008, pp. 103–110, 2020.
https://doi.org/10.1007/978-3-030-23884-1_14

a. Process
b. Product
c. Press
d. Person

The first dimension, process, is connected with the cognitive dimension of creativity that leads to new ideas. In a schematic representation this process can be divided in different steps: first of all the problem or the task is identified, thus triggering the following steps that include preparation, response generation and validation, the communication to others, at the end, the outcome. Creativity consists in following these steps in an unforeseen and unusual way, anyway leading to a new product.

Product is indeed the second dimension of creativity and includes material elements such as artifacts, artworks, innovative tools, as well as immaterial elements including theoretical perspective, new frame of reference, etc. Also the external environment, press, where creative process takes place, plays a relevant role.

The fourth dimension is person and refers to personality and cognitive traits that define a creative person. They include broad interests, independence of judgement, openness [13, 27], intrinsic motivation [3, 9], creative self-concept [5].

Moreover we can identify the so-called creative life-style that refers to originality in self-expression (e.g., designing ones own jewelry), interpersonal behaviour (e.g., creating a scrapbook of memories for a friend), culture participation (e.g., organizing a poetry recital), creative leisure activities of writing and visual arts (e.g., writing poetry, completing paintings), according to Ivcevic and Mayer [15]:

"Creativity also exists in everyday life; it permeates daily life in areas of self-expression and presentation, managing personal relationships, practical artistry, and culture participation".

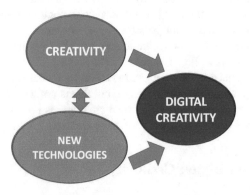

Fig. 1. The connection between creativity, new technologies and digital creativity

Creativity is therefore important, at individual level, as an individual resource [22], a way to adapt and as a latent power. In fact, literature highlighted that creativity could be interpreted as an individual resource, mostly in terms of adjustment. Indeed, creativity was considered one of the main personality traits useful both for adaptation and maladjustment of individuals to environment [10].

In Csikszentmihalyis terms, creative individuals are remarkable for their ability to adapt to almost any situation and to make do with whatever is at hand to reach their goals. Furthermore, Ogoemeka [16] (2011) found out that creativity was a potent predictor of social problem-solving and both creativity and cognitive ability were strong predictors of adjustment. Overall, creativity could have an important role in the development of individual and can be defined as an inherent latent power present in each person [4].

Educating to creativity is therefore a key challenge that educational context must face to allow individuals to exploit it. Moreover, the chance to use digital tools opens new possibilities to more and more people to be creative at different levels. Digital creativity is a fruitful way to promote and stimulate creativity, at the same time, giving value to digital tools in an inclusive framework promoting creativity and divergent thinking in everyone.

In the challenge that education is called to meet, a relevant role is played by teachers, as they are, together with students, the leading characters on the education stage. In education, creativity can stimulate imaginative activity generating outcomes that are original and valuable in relation to the learner [8] and digital technologies can be the medium to promote creativity, under the guidance of teachers that promote creative expression by the means of digital tools, teach how to use digital tools in a creative way, design, implement and propose to learners creative learning scenarios.

3   Approaches and Tools for Integrating Digital Creativity in Education

3.1   Game-Based Learning, Gamification and Serious Games

While playing a game, at every age, the player is involved in a challenge, cognitive or physical that leads her/him to learn something. This something can be the elements and the dynamic of the game itself or other contents that are connected to game scenarios, characters, history, etc.

We can say that playing a game implies learning and this feature can be exploited to make games an useful vehicle to transfer knowledge, as they provide challenges to player and force to acquire some abilities. This aspect can be recognized in every kind of game, such as playing cards, board games, role games; for the purposes of this lesson and for the amazing diffusion they have had in recent years, we will focus on digital games. Considering digital games, the comprehensive review by Tobias and colleagues on Game-based learning examines the empirical evidence on the effectiveness of using video and computer games to support learning process. It underlines that people do learn from games and that this process can be particularly effective if the game design process supports learners acquisition of the specific knowledge and skills the games mean to transfer and a good instructional design is integrated with the highly motivating features of games. In digital games for learning, 2 dimensions must be harmonized in order to reach educational results:

- the first dimension is related to psychology and education, in particular to the cognitive, motivational and affective dimensions of the learning process
- the second one is related to the use of ICT, information and communication technology and the technology serving educational purposes.

Technology is shaping the future of education: technologies like artificial intelligence, machine learning, and educational materials with a strong technological side (both software and hardware are recreating students learning processes, the role of educators and bringing revolution into classrooms. There are some features of games that are particularly relevant in producing positive effects on learning:

- MOTIVATION and ENGAGEMENT: Educational computer games are motivating and lead to "flow" experience that describes how smooth gaming results in deep engagement and, consequently, enhanced learning [17]. Related with this, important aspects of motivation including personal and game characteristics, usability issues and anticipated outcomes can enhance engagement.
- TIME ON TASK and LEARNING by DOING: The motivating aspects of game lead student to spend more time on task, an aspect that, since 80 s has been connected to better learning outcomes [23]. Moreover games allow to gain learning by doing, that is to say, to reach an experiential learning.

Experiential learning activities can help students to remain focused, as learning actively makes less likely to become bored; to learn differently, as they are more engaged emotionally and learn faster, because learning by doing requires problem-solving and critical thinking that accelerate learning and improve content retention.

Gamification is not game-based learning [20]. Indeed gamification is defined as follows: gamification is the application of game-design elements and game principles in non-game contexts. Gamification commonly employs game design elements to improve user engagement, organizational productivity, flow, learning, ease of use, usefulness of systems, physical exercise, traffic violations and more. Gamification consists in using game thinking and mechanics in a non-game context to engage users and it can be applied to almost every aspect of life.

We must also distinguish Game-based learning and Serious Games. In Game-based learning, games (traditional or digital) represent the main learning material. Serious games are games that have a goal which goes beyond entertainment in the direction of education and learning.

They can be both used in traditional educational setting, such as schools, or in extracurricular activities. A serious game or applied game is a game designed for a primary purpose other than pure entertainment. The "serious" adjective is generally prepended to refer to video games used by industries like defense, education, scientific exploration, health care, emergency management, city planning, engineering, and politics. Serious Games must be built following the following principles: Learning must not be incidental, on the contrary, teaching and learning should be intrinsic to game and educational game should follow the same game design process of entertainment games.

A nice example of this method is provided by Zombie Division, a serious game to learn arithmetic where the player is an ancient hero who faces skeletons wearing number. The player can uses several attacks, each one corresponding to a kind of division operated on the number linked to the target enemy. The player must take care of matching attacks with opponents through divisions, as enemies cannot be divided without the required attack.

Game-Based Learning, Gamification and Serious Games at School
At school games and Serious Games can be used:

- To design and build artifacts: for example Minecraft can be used as a tool to create models of cities, ancient cities, etc.
- To deliver contents about a certain area: for example, Pirates can be used to learn about history and geography of Caribbean
- To set and rule simulations: games are used to test theories and experiment in complex systems
- To get in touch with technology: games can be used to familiarize with technological tools, for example Robotics (Lego Mindstorms)
- To cover different roles: role playing games can be used to experiment different roles and identities
- To exercise at home
- To assess abilities and knowledge

3.2 Game-Based Learning in the DoCENT Project

In this section we will introduce the approach of Game-based learning (GBL), as it has been applied in the DoCENT project. We will start from the pedagogical underpinnings of GBL, we will go in more detail in SG application in DoCENT. Considering the pedagogical underpinnings, the GBL approach in DoCENT relies on these fundaments [1, 14]: student-centered methodologies, self-regulated learning processes, help establishing connections, exploration and discovery, immersion, safe environments that encourage risk-taking, flexible assessment strategies, collaboration, use a variety of sources, including ICT. Moreover Docent adopts a constructionist and participatory approach to GBL that allows to move from "playing to making", thus engaging students in creating and sharing digital games.

This is done promoting the adoption by teachers of manipulative technologies (Fig. 2), Educational robotics, game design and coding [29–31, 32]. A crucial point is how to integrate GBL in teachers activity. Indeed GBL can be applied in a very flexible way in educational context, following different pathways. It is possible to reformulate existing non-educational games for educational purposes.

Following this pathway, existing non-educational games are used in an educational context to promote the acquisition of specific knowledge, for example history, pollution and a wide range of subjects. In the case of MINECRAFT, the education edition is used by educators in grades K-12 to teach a range of subjects, from history and chemistry to sustainability and foreign languages, and can map lessons directly to specific learning outcomes and curriculum standards. It is possible to use existing educational games.

In this case, existing non-educational games are exploited to supported the acquisition of contents or the training of abilities. As an example, ENACT is a game, developed within the framework of the ENACT Project (a project funded by EACEA under Erasmus+ KA3 measure), that can be used to assess and train negotiation abilities through simulated interactions between the user and a virtual agent, controlled by an Artificial Intelligence system. Another chance is to design tailored games: if no existing game is for the educational goal to be achieved, it is possible to design a tailor made game, using editors such as Unity or Game Maker. This process, which allows to get a very customized tool, it is very demanding in terms of time and resources. It is also possible to exploit gamification. As explained before, gamification consists in using game-like elements in a wide variety of other context, including educational contexts, both formal and informal. The last possibilities is to build Serious Games that can be based on different computing approaches and can be used as teaching and learning tools and as assessment tools [6, 11, 12].

Fig. 2. Tangible materials that can be used in creative learning scenarios.

4   Conclusion and Future Directions

In this paper we have described the Docent project whose goal is to promote digital creativity in education context. Teachers' role is crucial in covering the gap between policies that value creativity and school practice that rewards convergent thinking. Digital technologies, games, tangible interfaces, robotics can be interesting and effective tools to integrate creativity in school curricula. The use of digital and physical materials, thanks to tangible interfaces, makes them very appealing for children and this aspect cannot be neglected in a school context where capturing attention is a continuous challenge. In next phases, the goal is to test the Docent methodology on wide sample and compare this methodology with more traditional ones, both in formal and informal contexts.

Acknowledgment. The DoCENT project (Digital Creativity ENhanced in Teacher education) is co-funded by the Erasmus+ programme of the European Union, in the call Key Activity 2 – Strategic Partnership and runs between October 2017 and September 2019.

References

1. Aldrich, C.: Learning by Doing: A Comprehensive Guide to Simulations, Computer Games, and Pedagogy in E-Learning and Other Educational Experiences. Wiley, Hoboken (2005)
2. Amabile, T.M.: Creativity in Context: Update to the Social Psychology of Creativity. Hachette, London (1996)
3. Amabile, T.M.: The social psychology of creativity: a componential conceptualization. J. Pers. Soc. Psychol. **45**(2), 357 (1983)
4. Baran, G., Erdogan, S., Cakmak, A.: A study on the relationship between six-year ability. Int. Educ. Stud. **4**, 105–111 (2011)
5. Barbot, B., Heuser, B.: Creativity and identity formation in adolescence: a developmental perspective. In: Karwowski, M., Kaufman, J. (eds.) The Creative Self. Academic Press, London (2017). ISBN 9780128097908
6. Cerrato, A., Ponticorvo, M.: Enhancing neuropsychological testing with gamification and tangible interfaces: the baking tray task. In: International Work-Conference on the Interplay Between Natural and Artificial Computation, pp. 147–156. Springer, Cham (2017)
7. Craft, A., Cremin, T., Burnard, P., Dragovic, T., Chappell, K.: Possibility thinking: culminative studies of an evidence-based concept driving creativity? Education 3-13 41(5), 538–556 (2013)
8. Cremin, T., Craft, A., Clack, J.: Creative little scientists: enabling creativity through science and mathematics in preschool and first years of primary education, D2. 2. Literature Review of Creativity in Education (2012)
9. Csikszentmihalyi, M.: Creativity, fulfillment and flow. TED (2008)
10. Csikszentmihalyi, M.: The creative personality. Psychol. Today **29**(4), 36–40 (1996)
11. Di Fuccio, R., Ponticorvo, M., Ferrara, F., Miglino, O.: Digital and multi-sensory storytelling: narration with smell, taste and touch. In: European Conference on Technology Enhanced Learning, pp. 509–512. Springer, Cham (2016)
12. Ferrara, F., Ponticorvo, M., Di Ferdinando, A., Miglino, O.: Tangible interfaces for cognitive assessment and training in children: LogicART. In: Smart Education and e-Learning 2016, pp. 329–338. Springer (2016)
13. Karwowski, M., Lebuda, I.: The big five, the huge two, and creative self-beliefs: a meta-analysis. Psychol. Aesthet. Creat. Arts **10**(2), 214 (2016)
14. Kolb, D.A.: Experiential Learning: Experience as the Source of Learning and Development. FT Press, Upper Saddle River (2014)
15. Ivcevic, Z., Mayer, J.D.: Creative types and personality. Imagin. Cogn. Pers. **26**, 65–86 (2006)
16. Ogoemeka, O.H.: Correlates of social problem-solving and adjustment among secondary school students in Ondo State Nigeria. Procedia Soc. Behav. Sci. **30**, 1598–1602 (2011)
17. Mattheiss, E.E., Kickmeier-Rust, M.D., Steiner, C.M., Albert, D.: Motivation in game-based learning: it's more than 'flow'. In: DeLFI Workshops, pp. 77–84 (2009)
18. Ponticorvo, M., Di Fuccio, R., Ferrara, F., Rega, A., Miglino, O.: Multi-sensory educational materials: five senses to learn. In: International Conference in Methodologies and Intelligent Systems for Technology Enhanced Learning, pp. 45–52. Springer, Cham (2018)

19. Ponticorvo, M., Di Ferdinando, A., Marocco, D., Miglino, O.: Bio-inspired computational algorithms in educational and serious games: some examples. In: European Conference on Technology Enhanced Learning, pp. 636–639. Springer, Cham (2016)
20. Ponticorvo, M., Rega, A., Miglino, O.: Toward tutoring systems inspired by applied behavioral analysis. In: International Conference on Intelligent Tutoring Systems, pp. 160–169. Springer, Cham (2018)
21. Ponticorvo, M., Rega, A., Di Ferdinando, A., Marocco, D., Miglino, O.: Approaches to embed bio-inspired computational algorithms in educational and serious games. In: CEUR Workshop Proceedings (2018)
22. Rhodes, M.: An analysis of creativity. Phi Delta Kappan **42**, 305–310 (1961)
23. Runco, M.A., Jaeger, G.J.: The standard definition of creativity. Creat. Res. J. **24**, 92–96 (2012)
24. Seaborn, K., Fels, D.I.: Gamification in theory and action: a survey. Int. J. Hum.-Comput. Stud. **74**, 14–31 (2015)
25. Sefton-Green, J., Brown, L.: Mapping learner progression into digital creativity (2014)
26. Sica, L.S., Ragozini, G., Di Palma, T., Aleni Sestito, L.: Creativity as identity skill? Late adolescents' management of identity, complexity and risk-taking. J. Creat. Behav. (2017). https://doi.org/10.1002/jocb.221
27. Stallings, J.: Allocated academic learning time revisited, or beyond time on task. Educ. Res. **9**(11), 11–16 (1980)
28. Sternberg, R.J., Lubart, T.I.: Investing in creativity. Am. Psychol. **51**(7), 677 (1996)
29. Tobias, S., Fletcher, J.D., Wind, A.P.: Game-based learning. In: Handbook of Research on Educational Communications and Technology, pp. 485–503. Springer, New York (2014)
30. Wands, B.: Digital Creativity: Techniques for Digital Media and the Internet. Wiley, New York (2002)
31. Werner, C.H., Tang, M., Kruse, J., Kaufman, J.C., Sprrle, M.: The Chinese version of the revised creativity domain questionnaire (CDQR): first evidence for its factorial validity and systematic association with the big five. J. Creat. Behav. **48**(4), 254–275 (2014)

Workshop on Student Assessment and Learning Design Evaluation in TEL Systems (TELAssess)

Workshop on Student Assessment and Learning Design Evaluation in TEL Systems (TELAssess)

Assessment in technology-rich environments is still considered an open issue. The scope of this workshop is twofold concerning both student assessment in TEL as well as evaluation approaches and tools that support the design of learning by educators usually at community level. Especially in the context of Learning Design (LD), student assessment and course evaluation are seen as an integral part of the LD cycle both in terms of conceptualization and in terms of actual implementation. Evaluation in TEL may concern various aspects such as student learning, educational content evaluation, and technology integration. Several tools and approaches have been proposed to support this evaluation. Innovative approaches that existing technological tools adopt, include reflection as a self-assessment or a community process providing means for analysing the learning process and/or outcomes, as well as for enhancing feedback provision and peer interaction.

The premise of the workshop is that opportunities provided by the integration of modern learning technologies along with effective LD approaches can change the ways we conceptualise and develop student assessments. The aim of the workshop is to strengthen the community on the topic and offer an arena for discussion, exchange of ideas and experiences among stakeholders.

The workshop focuses on practical approaches, methods and intelligent systems that promote student assessment and learning design evaluation in technology-rich environments. Contributions to this workshop may address diverse aspects related to assessment on technology-rich environments from various perspectives. We invite researchers and tutors who have an interest in the 21st century student assessment and learning design evaluation to contribute in the workshop in one of the following ways: demonstrate their novel in-progress or recent work including hands-on practice, identify challenges and report on non-successful initiatives and research we can learn from, or discuss empirical outcomes related to student assessment and learning design evaluation using TEL.

The topics of the workshop include (but are not limited to):

- The interplay of Learning Design and student assessment in TEL
- Approaches for learning design evaluation in TEL
- Tools that support learning design evaluation

- Approaches related to training for the creation of effective assessment in technology-rich environments
- Intelligent systems and assessment in TEL
- Active learning methods that incorporate assessment in TEL
- Design principles or patterns that can promote effective assessment in TEL
- The interplay of Learning Analytics and student assessment in TEL
- Alternative students' assessment approaches in TEL
- Innovative assessment approaches in MOOCS
- Assessment of 21st century student skills in TEL
- Peer and self-assessment in TEL

The format of the workshop includes interactive presentations and demonstrations.

Organiztion

Organizing Committee

Anna Mavroudi	Royal Institute of Technology, Sweden
Kyparisia Papanikolaou	School of Pedagogical & Technological Education, Greece
Elvira Popescu	University of Craiova, Romania

Program Committee

Maria-Iuliana Dascalu	Politehnica University of Bucharest, Romania
Giuliana Dettori	Institute for Educational Technology (ITD-CNR), Italy
Yannis Dimitriadis	University of Valladolid, Spain
Agoritsa Gogoulou	National and Kapodistrian University of Athens, Greece
Olga Fragkou	Hellenic Open University, Greece
Davinia Hernández-Leo	Universitat Pompeu Fabra, Spain
Yiannos Ioannou	Ministry of Education and Culture, Cyprus
Mohammad Khalil	University of Bergen, Norway
Zuzana Kubincova	Comenius University Bratislava, Slovakia
George Magoulas	Birkbeck University of London, UK
Katerina Makri	National and Kapodistrian University of Athens, Greece
Marco Temperini	Sapienza University of Rome, Italy
Panos Vlachopoulos	Macquarie University, Australia

Code Review at High School? Yes!

Zuzana Kubincová[✉] and Iveta Csicsolová

FMFI, Comenius University in Bratislava,
Mlynská dolina, 84248 Bratislava, Slovakia
kubincova@fmph.uniba.sk, ivetkacs@gmail.com

Abstract. Mutual evaluation of students' work is an educational strategy recently investigated by educational professionals. It can be implemented into the programming teaching in the form of code review. As supported by several studies, this activity can help substantially improve the program code and increase its development efficiency. Moreover, it can provide students with various benefits, such as soft skills development, streamlining their learning, promotion of social learning, and so on. In this paper, we present partial results from our research on using code review in informatics classes at secondary school.

Keywords: Programming · Code review · High school

1 Introduction

Peer review, a mutual evaluation of students' work, is one of the techniques that are recently in the center of attention of educational professionals [1, 2]. This activity can provide students with various benefits, such as improving their learning outcomes, development of their interpersonal skills and strengthening of social learning [3, 4].

One of the ways to introduce peer review in informatics teaching is to use this method to review program code in programming lessons – code review [5]. Such a learning activity is based on social constructivism [6] and is motivated by the extensive use of code review technique in software companies during the development of major software projects. As shown in many studies, code review in professional programmer teams leads to a significant improvement in the quality of the program code and the overall effectiveness of project development [7, 8].

Most of the published studies dealing with the code review in education were from university courses. In the context of Slovak research we were not able to find any relevant publications in this area related to the high school. That is why we started to carry out a study of educational activities using code review about a year ago. In this article we bring partial results of our research.

2 Code Review in High School Programming

Trying to employ code review technique in high school teaching and to determine its impact on student learning, we have designed activities related to this technique and tested them under ordinary teaching conditions.

© Springer Nature Switzerland AG 2020
E. Popescu et al. (Eds.): MIS4TEL 2019, AISC 1008, pp. 115–123, 2020.
https://doi.org/10.1007/978-3-030-23884-1_15

The research was conducted during the 2017/2018 school year in two classes of the third grade at a high school[1]. It was carried out in informatics classes (one hour a week), while programming in Python was taught.

Students have already programmed in Python in the previous school year, but they have not met code review. The previously taught themes were first reviewed and only afterward the teacher continued to teach new ones. Two types of code review activities were involved in teaching:

- Small reviews – short tests, each containing two short programs (only few lines of code) prepared by the teacher. In the first one, students tried to find out what would the program do if it was executed. In the second one, they should look for errors (syntactic and logical) in the given program and correct them (see [9]).
- Project reviews – mutual reviewing larger programs coded by students.

In this article, we focus mainly on the later activity – project reviews.

3   Code Review of Projects

After performing a few small reviews, which could be perceived as training, we let the students review each other's projects. In this review the students commented on the program code created by other students – an activity that was more similar to the real code review.

In the school year 2017/2018, we used code review of projects twice. In both cases, students developed the code at home as a homework. The first project was assigned to students to work out during the winter holidays, which was only a few weeks after the programming unit started to be taught. The entire period for the whole project activity including commenting on the others projects was 5 weeks. The second project was assigned to the end of the programming unit teaching (in April). Since in this project it was necessary to use all the knowledge gained during the whole period of programming teaching, it was considerably more difficult than the previous one. The students asked for a longer time to prepare the project, so the cumulative time they spent on it was 7 weeks.

In the first project, the students chose one of the three assignments – car racing, sketch application and postcard – so that in each group the project with the same input could occur several times. Each of these assignments has been specified in detail.

In the second project, the students were offered 17 different themes of assignments, such as Rock-Paper-Scissors, Lotto, Ball bouncing, Sophisticated sketch application, T-Rex, and more. They were asked to choose one of them, so that no theme occurred repeatedly in the same group of students. Detailed specifications for these project assignments were prescribed as well. Students could also suggest their own project theme which had to be approved by the teacher.

---

[1] Bilingual high school with 5-year study.

3.1   Participants

In the first half, 52 pupils participated in the study, while in the second half there were 55 of them. The students were divided into four groups, 3A1, 3A2, 3B1, and 3B2. In group 3A1 there were 13 students, of which there were three boys (in the second half plus one student); in group 3A2 there were 14 students, of which five boys. In group 3B1 there were 12 students (in the second half plus 1 student), of which there were four boys, and in group 3B2 13 students (in the second half plus 1 student), of which six boys.

3.2   Code Review Methodology

The code review of projects brought a new aspect to the project activity – a possibility to improve the project based on the comments received from the classmates. All project work could be divided into three phases involving the following steps:

A. Project development

- Selection of one of the proposed project themes
- Programming the project
- Submitting the project for reviewing

B. Reviewing a classmate's project

- Assigning of projects for review
- Commenting on the assigned project
- Submitting a commented project

C. Customization of the project

- Returning the commented project to its author
- Modification, completion, and/or improvement of the project
- Submitting the final version of the project for the teacher evaluation

The project development phase lasted three weeks for the first project, and up to five weeks for the second one (as a consequence of higher difficulty of the project).

Since we did not have any suitable tool for programs distribution and code reviewing, the students delivered their projects to the teacher via e-mail. She has assigned projects to other students for reviewing trying to guarantee the anonymity, so the project author did not know who reviewed her program, and the reviewer did not know, who was the author of the reviewed program.

The role of a student-reviewer was to check the assigned classmates' project: to find out whether she has followed all the parts of the assignment, to identify errors, to counsel what needs to be modified or added, and write her overall opinion on the project. There was one week reserved for this activity. Thereafter, the student e-mailed the commented program back to the teacher who returned it to its author.

In the final phase, which lasted for another week, the student's job was to incorporate the justified reviewer's comments into her program, or fix the mistakes she had found herself in the meantime. This way, students have been given the opportunity to correct the deficiencies in the project and improve it in order to achieve a better mark.

The teacher monitored the students' activity and the results of their work at each phase (coding as well as commenting). After the last phase, she sent an individual assessment to each student, where she commented on the student's work during particular phases of the project.

3.3   Results

Project Evaluation Results

For both projects, the specific requirements that needed to be met were specified for each project theme. The common criteria that were taken into account in the final project evaluation by the teacher were as follows: full functionality, correctness and authenticity of the project, commenting on the classmate's project (finding errors, proposing solutions, overall view of the project), and customization of the own project according to the classmate's comments. For project features beyond the assignment, students could earn additional bonus points. As both projects were challenging tasks for students, they have contributed more to their overall evaluation in informatics.

The results of the first project were very satisfactory. After receiving comments from classmates, the students have corrected many bugs and achieved a very good rating from the teacher. Up to 73% of projects achieved a percentage rating corresponding to the mark "excellent". More than one third of all projects were rated by the teacher as exceptionally good, exceeding her expectations.

As the second project was considerably more challenging for the students than the first one, the distribution of students programs in particular categories of evaluation was more even than in the case of the first project. Nevertheless, almost 55% of students achieved the best rating, and again almost a third of the projects were identified as exceptionally good by the teacher.

Analysis of the Questionnaire Results

After the evaluation of each of two projects by the teacher, the students were asked to anonymously fill out a questionnaire. The goal of this survey was to find out their views on mutual review of projects and their experience of this activity. The questionnaires were broadly focused, but here we will only deal with code-review issues.

The first questionnaire was filled out by 43 students and the second one by 44 students of which in both cases approximately 61% of girls and 39% of boys (±0.5%).

Since before the first project the students did not have any experience of reviewing a program code longer than a few lines (small reviews), we were interested in their opinion on the complexity of commenting on the project. We also kept the question in the questionnaire for the second project. Although the distribution of responses in the second questionnaire was not as explicit as in the first one (Fig. 1), the share of students who did not consider the commenting on the classmate's project as challenging (values 1–3) was very high and roughly the same in both cases (90.8% and 88.6%).

The teacher also confirmed that in both projects, the students were capable to understand somebody else's code, to find errors in it, and to comment on it. It has even happened that they found an error that the teacher did not notice. In the first project, only a few of the students were unable to detect most of the errors in the classmate's program. Many have been able to find all the mistakes. In the second project, due to its higher difficulty, some students were unable to advice how to fix the program, but many of them managed it, which the teacher appreciated when giving them feedback.

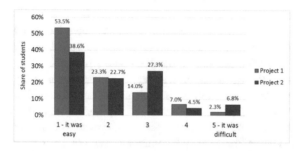

Fig. 1. The students' answers to the question "Was it difficult to comment on the classmate's project?" Scale from 1 (it was easy) to 5 (it was difficult)

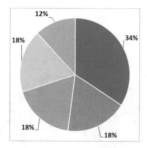

Fig. 2. The students' answers to the question "How difficult was it to comment on this project compared to the commenting on the previous project?" ■ It was easier just because we already commented on the project, ■ It was easier because I am better in programming and better understand the code now than in the first project, ■ It was as difficult as in the previous project, ■ It was more difficult because the project was harder, ■ Other

In the second questionnaire, we asked the students to compare the difficulty of commenting on the first and second projects (Fig. 2). For 52% of students, it was easier to do, either because they already commented on the project (34%) or because they thought they were better programmers now and better understood the code than in the first project (18%). The share of students who perceived the commenting now more difficult than before was 18%, and they rationalized it by the harder project. 12% of students added their own response – with about half of them stating it was easier to comment on this project and it was more difficult for the other half. Overall, commenting on the heavier project was easier for nearly 60% of students and it was harder for less than a quarter of students.

We also checked what problems the students encountered when commenting on the project (Fig. 3). In the first project, up to 63% of them said they had no problem with commenting, 16% of students could not get oriented in the classmate's code, 16% did not comment because they did not find any errors. For the second project, students' responses were distributed into more categories. Also such responses appeared here that have not appeared in the previous questionnaire (e.g. "I did not understand the project at all" or "I was not able to describe the error I have found"). However, despite the fact that the students described the second project as very demanding, 55% of them had no problem with commenting on the classmate's program.

The answers to the question of whether the classmate's comments helped the student when completing the project were also important for us. The responses were quite different even in the same questionnaire (Fig. 4). In the first project, more than half of students declared positive experience of classmate's comments and only 9% of students responded that the received comments did not help them. In the second project, students' responses were almost completely evenly distributed into all categories. Apparently, due to the difficulty of the task, the share of students who failed to clearly and meaningfully comment on their classmate's program rose to 23%. However, the amount of comments rated positively (39%) was higher than those rated negatively.

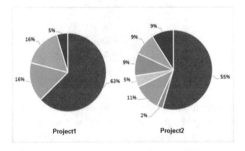

Fig. 3. The students' answers to the question "What caused you a problem when commenting on a classmate's project?" ■ I did not have any problems in commenting on the project, ■ I did not understand the project at all, ■ I was not able to orientate in the program well, ■ I was not capable of finding out why project is not working, ■ I was not able to describe the error I have found in the project, ■ There was nothing to comment as I did not find any errors, ■ Other

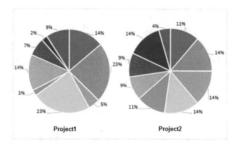

Fig. 4. The students' answers to the question "Did the comments from your classmate help you when finishing the project?" ■ Yes, I got good comments that helped me a lot. ■ Yes, I received good comments that helped me, and I was delighted to be praised for the project. ■ I received good comments but did not know how to integrate them into the project. ■ Some comments helped me, some did not. ■ My project worked, my classmate found no mistakes, so I did not get any comments that I could work with. ■ The reviewer did not find any mistakes, but I was pleased that he praised my project. ■ No, they did not help me because I did not understand them and did not know how to integrate them into the project. ■ No, I have received nonsense comments that have not helped me at all. ■ Other

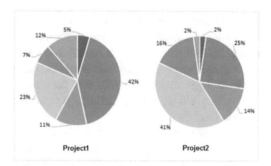

Fig. 5. The students' answers to the question "What were the benefits of the comments you received from the classmate?" ■ They helped me better understand the project assignment. ■ They helped me to fix errors I did not notice before the first submission of the project. ■They helped me to fix errors I failed to fix before the first submission of the project . ■They did not help me at all. ■ I have not received any relevant comments. ■ Other

In both projects, the students saw the contribution of comments (Fig. 5) from their classmates mainly in helping them to fix errors they did not notice or failed to fix them before the first submission of the project, or helped them some other way (58% for the first and 41% for the second project). The higher difficulty of the second project seems to have influenced the answers to this question too – the proportion of students who declared that the comments did not help them increased from 23% to 41%.

We also asked the students whether they had gained something by commenting on the classmate's project. In both questionnaires, the most frequent answer (37.2% and 27.3%) was that they were better aware of various mistakes that might occur in the program. Further, the students often indicated that they learned how the classmate understood the assignment, gained experience with code reviewing, and learned to give a constructive critique. In both cases, roughly the same share of students (16.9% and 15.9%) said that they did not gain any profit from commenting on the project.

4   Conclusions

By using code review in programming, we are trying to help students gain better insight into the programming code, learn to detect bugs in the program, and understand different approaches to solving the same problem.

We tested the code review activities with students on two projects with a time span of several months. The first project was fairly simple and the students achieved very satisfactory results possibly also thanks to the benefits from comments. The second project was considerably more challenging thanks to new, more complex programming concepts that students had to meaningfully combine, but also due to the more complex assignment. Also, commenting on the second project was more challenging for students. The results of this project better divided students into groups according to the knowledge of programming. However, the overall results of this project were also very good.

From the beginning, the code review activity was not perceived utterly positively by students since they saw it as an extra work. When commenting on projects, students had to understand a program that had a different specification than their own project. Thus they considered this task to be complex and time-consuming. Sometimes they mentioned as a problem a large number of mistakes they have found and had to comment that way so that they did not tell the author the correct solution at once, but only guide her to the right direction.

However, many learners have realized the usefulness of such a way of programming learning, and have stated, for example, that it also made them to think about the programming problems they did not encounter in their own project. In addition, many students also appreciated the feedback from the teacher who helped them to understand exactly what they did well and badly in the project, and so they were given the chance to improve their work later on.

The students were expected to submit already in the first phase nearly completed project. From a teacher's point of view, however, the students in most cases handed in much better projects after correcting the errors found by their classmates or when the students themselves checked their projects after some time span. By studying both classmate's, and their own code, students learned to prevent errors, as well as to find and correct them. Before using the code review activity in teaching these classes students often asked the teacher for help even because of small error messages or syntax errors in their programs. This activity has taught them some self-sufficiency while reading the code.

Based on the results of our research, we can presume that the code review can be used in teaching at secondary school. Students learn to read, navigate, and understand someone else's code, not only their own. Their algorithmic thinking is also enriched by seeking alternative solutions or finding an alternative solution to the problem in the code of somebody else. Code review activity in high school teaching brings a lot of benefits to students, although they often see it as an extra work.

For the teacher, the code review brings a different view of the student and the possibility to better track her progress in programming and algorithmic thinking.

Nevertheless, it also brings obstacles. Submitting projects in several phases, redistributing them for review and forwarding them back to the authors without any software tool is a time-consuming work for the teacher. Nevertheless, we think that the efforts that teachers and students have to make when integrating the code review into teaching are sufficiently balanced by the benefits that such an educational strategy brings to them.

Acknowledgement. This work was supported from Slovak national project VEGA 1/0797/18.

References

1. Purchase, H., Hamer, J.: Perspectives on peer-review: eight years of Aropä. Assess. Eval. High. Educ. **43**(3), 473–487 (2018)
2. Dropčová, V., Kubincová, Z.: Team-based projects and peer assessment. IT works!. In: International Conference on Interactive Collaborative Learning, pp. 112–127. Springer (2016)
3. Tseng, S.C., Tsai, C.C.: On-line peer assessment and the role of the peer feedback: a study of high school computer course. Comput. Educ. **49**(4), 1161–1174 (2007)
4. Topping, K.J.: Peer assessment between students in colleges and universities. Rev. Educ. Res. **68**, 249–276 (1998)
5. Kubincová, Z., Homola, M.: Code review in computer science courses: take one. In: International Conference on Web-Based Learning, pp. 125–135. Springer (2017)
6. Falchikov, N., Goldfinch, J.: Student peer assessment in higher education: a meta-analysis comparing peer and teacher marks. Rev. Educ. Res. **70**, 287–322 (2000)
7. Boehm, B.W., et al.: Software Engineering Economics. Prentice-Hall, Englewood Cliffs (1981)
8. Wiegers, K.E.: Peer Reviews in Software: A Practical Guide. Addison-Wesley, Boston (2002)
9. Kubincová, Z., Csicsolová, I.: Code review in high school programming. In: 17th International Conference on Information Technology Based Higher Education and Training (ITHET), pp. 1–4. IEEE (2018)

Exploring the Peer Assessment Process Supported by the Enhanced Moodle Workshop in a Computer Programming Course

Gabriel Badea[1], Elvira Popescu[1], Andrea Sterbini[2],
and Marco Temperini[2(✉)]

[1] Computers and Information Technology Department, University of Craiova,
Craiova, Romania
[2] Computer, Control, and Management Engineering Department,
Sapienza University, Rome, Italy
marte@diag.uniroma1.it

Abstract. Supporting peer assessment in learning management systems is an important educational issue. The widespread Moodle platform relies on a plugin called Workshop for providing such peer evaluation functionality. In a previous work, we proposed an extension of the plugin with student modeling capabilities, based on a Bayesian Network approach. In the current paper we aim to experimentally validate this *Enhanced Workshop* module, by using it in the context of an Introduction to Computer Programming course. An experience report of the peer assessment process is provided, focusing on the support offered by the module. The results are also analyzed, exploring the relationship between student models and grades.

Keywords: Peer evaluation · Moodle · Bayesian network model ·
Student model

1 Introduction

Peer assessment (PA) has an important place in education, stimulating students' critical thinking and evaluation abilities and encouraging peer learning [6, 8, 10]. Teachers may also benefit from implementing PA in their classes, by sharing the evaluation tasks with students and thus reducing their assessment workload. In recent years, several online platforms have been proposed to support the PA process, such as: CrowdGrader [1], PeerStudio [5], CaptainTeach [9] or Mechanical TA [12].

Apart from the various standalone platforms, peer assessment is also used in learning management systems. *Moodle* [7], for example, provides an approach to hold PA sessions through the dedicated *Workshop plugin* [11]. Starting from this plugin, in [3] we proposed an enhanced version, which provides student modeling capabilities, based on a Bayesian Network (BN) approach. More specifically, the *Enhanced Workshop (EW)* module uses a Bayesian Network Service (BNS) for computing the student model, as described in [2]. In addition, *EW* provides various metrics regarding

© Springer Nature Switzerland AG 2020
E. Popescu et al. (Eds.): MIS4TEL 2019, AISC 1008, pp. 124–131, 2020.
https://doi.org/10.1007/978-3-030-23884-1_16

the reliability of the computed models, as well as an enhanced interface with suggestive graphical visualizations for both teachers and students.

In this paper we describe and analyze the pilot use of EW in the context of an Introduction to Computer Programming course, involving 32 undergraduate students. Five PA sessions were performed, for which EW collected the student answers, the peer evaluations, and the teacher's marks; the final exam grades obtained by the students are also available. The goal of the paper is to provide an experience report of the peer assessment process, describing the unfolding of the PA sessions and the support offered by EW. We also analyze the results, investigating the quality and usefulness of the student models computed by EW.

We start by providing some background information about the Bayesian Network Service for PA, and a brief description of its integration in EW (Sect. 2). Next, we describe the context of study, outlining the PA settings in the IPC course (Sect. 3). Subsequently, we analyze the reports provided by EW (Sect. 4) and explore the relationship between student models and grades (Sect. 5). Finally, we draw some conclusions and future work directions (Sect. 6).

2 Background on BN Modeling and Its Integration in EW

In what follows, we provide a brief description of the Bayesian Network approach for student modeling based on peer assessment data, as presented in [4]. In the BN representation of data coming from a PA session, a student, her/his assessments and model are rendered as a set of discrete variables of the network. Each variable provides, normally as a distribution of probability: (1) the student's level of knowledge (competence) about the topic of the question proposed in the session (variable K); (2) the quality of the assessments provided by this student (i.e., her/his judgment skills - variable J); (3) the correctness of the student's answer, expressed as a grade (variable C), and (4) the peer assessments received by the student's answer (a variable G for each received grade). By propagation of the Gs and of the grades given by the teacher on a subset of the answers (if available), the other variables are computed. The BN assumes dependencies among the variables, given by conditional probability tables (CPT), such that G depends on J (of the grading peer) and C (of the graded peer), and both J and C depend on K. More details can be found in [4].

Based on the above framework, a Bayesian Network Service was proposed in [2], which was subsequently integrated in EW, as described in [3]. In brief, for each session, EW sends to the BNS the following parameters: (1) the *PA data* (the grades given by students); (2) optionally, the *K values* of the current student models; (3) optionally, a new *teacher grade*, to be used for network propagation; (4) optionally, several other parameters for setting the algorithms to be used for C computation, CPT and network variables initialization etc. After invocation, the BNS returns the newly computed *K and J values* (i.e., the student model updates), the newly computed *C values*, and a suggestion for the most convenient answer to be graded by the teacher next.

In addition, EW also provides a measure of *reliability* for the student models computed by the system. Reliability indicates the amount of trust the teacher can confide in the model's values, in order to make decisions based on it (such as assignment of further work, remedial activities, or even final exam grade). The number of sessions in which a student participates and the variability of her/his performance in different sessions impact the system's ability to compute the student model. Hence, we defined the *reliability* of a model as a function of the *continuity* shown by the student in participating in sessions and of the *stability* of her/his performance along the sessions; more details can be found in [3].

3   Study Context and Unfolding

EW was used in an experimental activity involving 5 optional PA sessions in a class of Introduction to Computer Programming (ICP), undergraduate level, Computer Engineering study program. 32 students (27 males, 5 females) were involved in the study. The course was split over two semesters (2017–2018 academic year) and the PA sessions were held as follows: two in December, one in April, one in May, and one in June. We experienced a decreasing number of participants along the sessions (23, 21, 14, 9, and 7 respectively), partly due to the long interval between the second and the third sessions and to contextual conditions (other heavily demanding courses proposing unplanned intermediate exams). More specifically, the involvement was as follows: 13 students participated in only one session, 9 in two sessions, 3 in three sessions, 1 in four sessions, and 6 in all the five sessions held.

The students were registered on our experimental Moodle instance, and were ensured access to the PA sessions in due time. Each session was comprised of a question, three assessment criteria, and two main phases: submission and assessment.

During the *Submission Phase* (about 70–90 min long), the student was requested to provide a textual answer to the assignment. The question regarded a programming problem, involving concepts that the students should have studied and trained upon over the past three weeks. The answer was supposed to provide an algorithm for solving the problem and programming code to implement the algorithm. The assessment criteria (and their weights) were also posted in this phase, along with the question, in order to provide guidance about the relevant aspects to be covered by the answer.

During the *Assessment Phase* (about 50 min), each student was assigned three peers' answers to evaluate. The student was supposed to provide three scores (on a 1 to 100 scale), one for each assessment criterion. The overall submission grade of an answer was computed as a weighted sum of these three scores. The criteria were referring to the quality of the algorithm, effectiveness of the implementation, proper formatting, readability and commenting of the code. An example of one of the prescribed criteria is as follows: *"Is the algorithm clear and correct? Are the steps clearly stated and consistent with one another? Does the algorithm actually solve the problem? Out of the 100 available points, use 60 to evaluate correctness, and 40 to appraise clearness and organization of the algorithm. [weight: 30%]"*.

Two additional phases are included in EW, which are of interest for the teacher only: *Setup Phase* and *Grading Evaluation Phase*. During the former, the teacher configures the EW session, specifying its components, weights, and establishing access constraints. During the latter, the teacher provides grades for students' answers, in addition to the ones provided by the peers. While this is optional, in our ICP course the teacher chose to grade all answers himself, in order to provide a reference point for the analysis, as discussed in the next sections.

4 Insights from EW Reports and Charts

In what follows we present some of the peer assessment and student model reports provided by EW for the teacher and the students of the ICP course.

For each session, the teacher can visualize a *workshop grading report*, displaying the session data for every student: *answer grade* (computed from the C value), *competence* (K) and *assessment capability* (J). For the student model variables (K and J) both "single session" and "cumulated" values are shown. The former are computed by EW based on current session data only, while the latter are based on all previous PA sessions, aiming to offer a more comprehensive outlook on students' performance. In addition, all individual peer assessments are visible (and linked) in this table. The report for the last session of our ICP course can be seen in Fig. 1; red color is used for lower values and green color for higher values, to easily highlight students' proficiency levels.

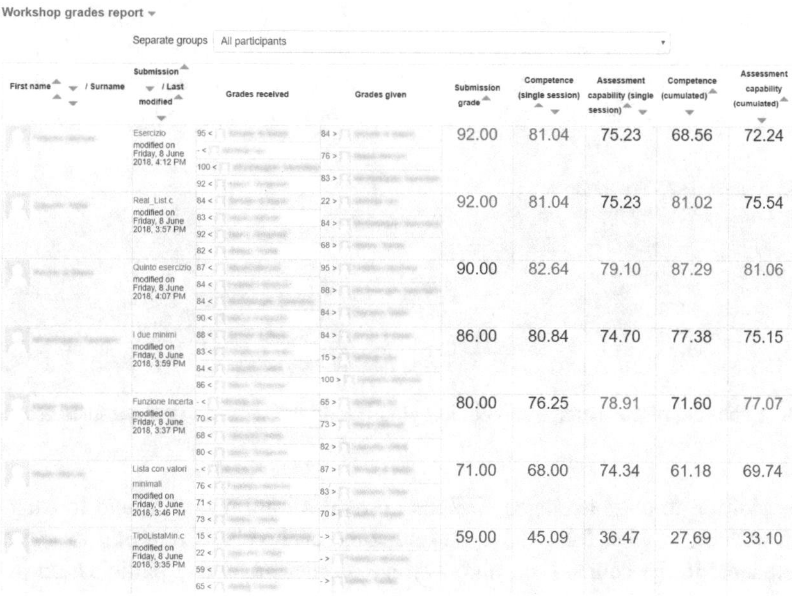
Workshop grades report

Separate groups: All participants

First name / Surname	Submission / Last modified	Grades received	Grades given	Submission grade	Competence (single session)	Assessment capability (single session)	Competence (cumulated)	Assessment capability (cumulated)
	Esercizio modified on Friday, 8 June 2018, 4:12 PM	95 < - < 100 < 92 <	84 > 76 > 83 >	92.00	81.04	75.23	68.56	72.24
	Real_List.c modified on Friday, 8 June 2018, 3:57 PM	84 < 83 < 92 < 82 <	22 > 84 > 68 >	92.00	81.04	75.23	81.02	75.54
	Quinto esercizio modified on Friday, 8 June 2018, 4:07 PM	87 < 84 < 84 < 90 <	95 > 88 > 84 >	90.00	82.64	79.10	87.29	81.06
	I due minimi modified on Friday, 8 June 2018, 3:59 PM	88 < 83 < 84 < 86 <	84 > 15 > 100 >	86.00	80.84	74.70	77.38	75.15
	Funzione Incerta modified on Friday, 8 June 2018, 3:37 PM	- < 70 < 68 < 80 <	65 > 73 > 82 >	80.00	76.25	78.91	71.60	77.07
	Lista con valori minimali modified on Friday, 8 June 2018, 3:46 PM	- < 76 < 71 < 73 <	87 > 83 > 70 >	71.00	68.00	74.34	61.18	69.74
	TipoListaMin.c modified on Friday, 8 June 2018, 3:35 PM	15 < 22 < 59 < 65 <	- > - > - >	59.00	45.09	36.47	27.69	33.10

Fig. 1. EW grades report for the 5th (last) session (teacher interface, session level)

In addition, at a more general level (*course level*) the teacher can visualize the aggregated data related to the sequence of sessions held in a course. In this *general student models grades report*, data representative of the progression along the sessions are shown for each student: average submission grade, cumulated values for K and J (the evolution of the model in time), and reliability of the model, with both its components, continuity and stability. An excerpt from this report, including the first 10 students, ordered according to their average submission grade at the end of the course, is presented in Fig. 2. It should be noted that this report includes all students who participated in at least one session, and the average submission grade only takes into account the marks for the submitted solutions (without penalizing for missed assignments); nevertheless, the *Continuity* column accounts also for the missing submissions, providing a more comprehensive perspective on students' work.

General student models grades report ▾

Name	Submission Grade	Competence	Assessment Capability	Continuity	Stability	Reliability
	95.00	79.94	63.58	20.00	53.46	33.38
	94.00	90.73	81.62	66.67	62.22	64.89
	86.80	87.29	81.06	100.00	92.70	97.08
	86.20	81.02	75.54	100.00	90.61	96.24
	84.80	68.56	72.24	100.00	89.37	95.75
	82.67	77.38	75.15	73.33	59.44	67.78
	82.50	74.67	74.80	20.00	59.37	35.75
	81.00	72.35	68.59	6.67	67.60	31.04
	80.00	70.07	62.08	6.67	68.00	31.20
	77.00	78.03	83.29	20.00	69.20	39.68

Fig. 2. EW general student models grades report (teacher interface, course level)

The instructor can also visualize the evolution of the average grade over the sessions, which is useful for spotting differences in the difficulty of the assignments. As seen in Fig. 3, the average grade decreases slightly over the first 3 sessions and then increases in sessions 4 and 5. This can be explained by the fact that fewer students attended these last two sessions, likely the ones with a higher level of knowledge and motivation.

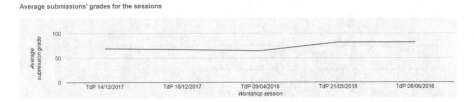

Fig. 3. Evolution of the average submission grade along the semester (teacher interface, course level)

In addition to class averages, EW also provides information related to each individual student. In what follows we present some of the charts referring to one of the best students in the course (i.e., high average teacher grade and participation in all 5 sessions), denoted *StudentA*. Figure 4 presents StudentA's results after the third session; we can notice that his ranking within the class is very good (solid green bar), and he has also improved his performance compared to the last session (hatched green bar). Figure 5 depicts the evolution of StudentA's performance during the course; it can be seen that his competence level is slightly increasing from one session to the other, relatively similar to the submission grade; his assessment capability score varies to a larger extent (reaching a maximum value in session 3 and a minimum value in session 4), but it is still constantly above average.

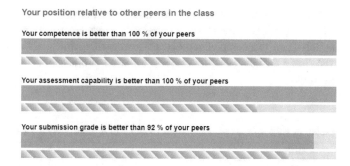

Fig. 4. StudentA's performance visualization for session 3 (student interface, session level)

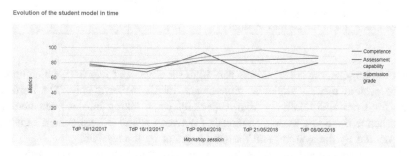

Fig. 5. StudentA's evolution of scores over the sessions (student interface, course level)

5   Insights from the Student Models and Grades

In what follows we aim to assess the consistency of the student models computed by EW, by comparing them with the actual student performance (i.e., grades awarded by the teacher for each assignment). The average values and standard deviation for each session are included in Table 1. As mentioned also in the previous section, the increase in grades for sessions 4 and 5 can be explained by the significantly lower number of attending students (presumably those with higher skills and motivation).

Table 1. Teacher grades, K and J values for each session (average and standard deviation)

Session	S1	S2	S3	S4	S5
Teacher grade	AV: 69.13	AV: 67.14	AV: 64.50	AV: 81.67	AV: 81.43
	SD: 19.19	SD: 12.81	SD: 22.51	SD: 15.68	SD: 11.54
K single	AV: 61.87	AV: 58.01	AV: 60.81	AV: 70.00	AV: 73.56
	SD: 18.97	SD: 15.36	SD: 20.89	SD: 12.28	SD: 12.51
J single	AV: 63.50	AV: 57.20	AV: 64.20	AV: 60.31	AV: 70.57
	SD: 15.33	SD: 14.04	SD: 17.50	SD: 9.39	SD: 14.04

Given the limited number of students, we could not perform in-depth statistical analyses. Instead, we computed the Pearson correlations between the K and J values of the student models (i.e., the competence and assessment capability estimated by the system) and the grade assigned by the teacher for each submission. The results are summarized in Table 2.

Table 2. Correlation values between the student model and teacher grades for each session

Session	S1	S2	S3	S4	S5
Correlation between *K single* and *Teacher grade*	0.98	0.90	0.90	0.98	0.95
Correlation between *K cumulated* and *Teacher average grade* (up to the current session)	N/A	0.95	0.94	0.98	0.93
Correlation between *J single* and *Teacher grade*	0.89	0.58	0.72	0.64	0.80
Correlation between *J cumulated* and *Teacher average grade* (up to the current session)	N/A	0.67	0.82	0.93	0.89

As can be seen, the correlation is very high for K values (over 0.9), which indicates that the system models students' competence in a consistent way. The correlation values for J are also relatively high, especially when taking into account cumulated values; hence there is a strong relationship between students' assessment capabilities (as computed by the system) and the grades given by the teacher.

We were also interested to explore the relationship between the assignment performance and the final exam grades obtained by the students. The exam has both a written and an oral component and its grade does not take into account the students' assignment work during the semester. According to the university regulations, the students can sit for the exam any time after the end of the course; so despite the fact that the course ended in June 2018, only 19 students of the 32 participating in the experiment have taken the exam so far (up to January 2019). Hence in what follows we only consider the data collected from these 19 students. Their involvement with the peer assessment sessions is diverse: 5 students attended all 5 sessions, 1 student attended 4 sessions, 3 students attended 3 sessions, 5 students attended 2 sessions and 5 students attended only 1 session.

In this case, the correlation was less strong: 0.49 between K and exam grade, 0.68 between J and exam grade. This can be explained by the fact that the assignments were optional and were not a part of the final grade; hence some students chose to skip them or not put in sufficient effort. Indeed, when taking into account the *Reliability* score (and especially the *Continuity* component), the correlations seem to increase for higher reliability values. However, given the limited sample size, adequate statistical analysis tests could not be performed; hence, more experimental data is needed to investigate the influence of the *Reliability* score.

6 Conclusion

The paper described the pilot use of the Enhanced Workshop plugin in Moodle, in the context of an Introduction to Computer Programming course. 32 students successfully used the system for assessing their peers' programming assignments. The reports and charts provided by the plugin brought helpful insights both for students and teachers. Strong correlations were obtained between the students' competence and assessment capability (computed by EW) and the assignment grades given by the teacher. Medium correlations were also found with the final exam grades, which indicates that the student models computed by the system may be used for early prediction and intervention during a course. This is especially the case for students who attended all or most of the peer assessment sessions, as the reliability of the models appears to be higher.

While the results are promising, the number of students involved in our study was limited, so further analyses are needed to investigate the reliability issue. We therefore plan to use EW in future courses, with a higher number of students and peer assessment sessions.

References

1. Alfaro, L., Shavlovsky, M.: CrowdGrader: a tool for crowdsourcing the evaluation of homework assignments. In: Proceedings of SIGCSE 2014, pp. 415–420. ACM (2014)
2. Badea, G., Popescu, E., Sterbini, A., Temperini, M.: A service-oriented architecture for student modeling in peer assessment environments. In: Proceedings of SETE 2018. LNCS 11284, pp. 32–37. Springer (2018)
3. Badea, G., Popescu, E., Sterbini, A., Temperini, M.: Integrating enhanced peer assessment features in moodle learning management system. In: Proceedings ICSLE 2019. Lecture Notes in Educational Technology, pp. 135–144. Springer (2019)
4. De Marsico, M., Sciarrone, F., Sterbini, A., Temperini, M.: Supporting mediated peer-evaluation to grade answers to open-ended questions. EURASIA J. Math. Sci. Technol. Educ. **13**(4), 1085–1106 (2017)
5. Kulkarni, C., Socher, R., Bernstein, M.S., Klemmer, S.R.: Scaling short-answer grading by combining peer assessment with algorithmic scoring. In: Proceedings of L@S 2014. pp. 99–108. ACM Press (2014)
6. Liu, N., Carless, D.: Peer feedback: the learning element of peer assessment. Teach. High. Educ. **11**(3), 279–290 (2006)
7. Moodle Learning Management System. https://moodle.org/. Accessed 07 Feb 2019
8. Pearce, J., Mulder, R., Baik, C.: Involving students in peer review. case studies and practical strategies for university teaching. Centre for the Study of Higher Education, University of Melbourne (2009)
9. Politz, J.G., Patterson, D., Krishnamurthi, S., Fisler, K.: CaptainTeach: multi-stage, in-flow peer review for programming assignments. In: Proceedings of ITiCSE 2014, pp. 267–272 (2014)
10. Topping, K., Smith, E.F., Swanson, I., Elliot, A.: Formative peer assessment of academic writing between postgraduate students. Assess. Eval. High. Educ. **25**(2), 149–169 (2000)
11. Workshop plugin. https://docs.moodle.org/23/en/Workshop_module. Accessed 07 Feb 2019
12. Wright, J., Thornton, C., Leyton-Brown, K.: Mechanical TA: partially automated high-stakes peer grading. In: Proceedings of SIGCSE 2015, pp. 96–101. ACM (2015)

Reducing Free Riding: CLASS – A System for Collaborative Learning Assessment

Olga Viberg[(⊠)], Anna Mavroudi, Ylva Fernaeus, Cristian Bogdan, and Jarmo Laaksolahti

School of Electrical Engineering and Computer Science,
KTH Royal Institute of Technology, 10044 Stockholm, Sweden
{oviberg,amav,fernaeus,cristi,jarmola}@kth.se

Abstract. In today's era of digitalization of education, Computer Supported Collaborative Learning is becoming increasingly important in higher education. This type of learning has been frequently associated in the recent research literature with student regulation, feedback from peers and a student assessment schema which can incorporate both formative and summative assessment strategies. This work-in-progress paper presents the CLASS system which caters for all these aspects. Furthermore, the system supports mechanisms for the prevention of the free riding phenomenon, which has been reported in the literature as one of the most important disadvantages in group student work. The paper discusses the higher education context in which the CLASS system was developed and used, along with its design affordances and how these affordances can facilitate CSCL. The paper can be useful to designers and developers of CSCL systems as well as to practitioners that are interested in how they can exploit CSCL with their students working in groups.

Keywords: Computer Supported Collaborative Learning · Assessment · Free riding · Higher education

1 Introduction

Computer-Supported Collaborative Learning (CSCL) involves social knowledge building supported by some technological environment(s). CSCL, if implemented appropriately, can provide "an ideal environment in which integration among students plays a central role in the learning process" (Roberts 2005, p.2). Typically, CSCL tools are designed to facilitate the conditions that harness learners' processes of knowledge co-construction. As suggested by Kobbe et al. (2007), computer-supported collaboration scripts - that aim at fostering collaborative learning by shaping the way that learners interact with one another - carry the benefit of reducing the coordinative effort both on the teachers' and the students' part.

Even if the success of CSCL heavily depends on whether students function properly in groups and can overcome barriers, such as Free Riding (FR), there is a lack of research specifically for the FR phenomenon (Slof et al. 2016). Although FR is the

© Springer Nature Switzerland AG 2020
E. Popescu et al. (Eds.): MIS4TEL 2019, AISC 1008, pp. 132–138, 2020.
https://doi.org/10.1007/978-3-030-23884-1_17

most cited disadvantage of group work and while the effects of interpersonal relationship in CSCL have recently started to attract the interest of the technology-enhanced learning (TEL) community, there are still limited studies that focus FR in the context of higher education (El Massah 2018). To compensate for this lack, this paper presents a *work-in-progress* study that exploits the affordances of the CLASS (Collaborative Learning Assessment) system showcasing how it facilitates student individual learning through processes of self-reflection, the provision of student feedback and ultimately, how it impedes FR.

2 Background

El Massah (2018) defines *free-riders* as team members who lack commitment to the group goals while reaping the benefits of the group's achievements. Consequently, they can impede the success of collaborative learning and CSCL. To overcome FR in collaborative learning settings, researchers found that adding an element of risk, namely a rotation of group members reduces the free-rider problem and reflects a significant positive impact on student performance, measured by their grades (Joyce 1999). Moreover, to reduce FR and to promote active learning, Swaray (2012) suggested an evaluation of a group project designed. The levels of the group members' contributions were supervised and the assessment method received input from the students' final presentation, the report, a short answer in the discussion and their reflective practice.

With respect to studies on group dynamics that affect CSCL processes, Slof et al. (2016) highlight the effects of agency and communion in the group, whereas Chavez and Romero (2012) focus on group awareness. Others emphasize group regulation and social-emotional interactions (Kwon et al. 2014). In addition, such characteristics as positive interdependence, individual and group accountability have been discussed (Gress et al. 2010).

The contribution of the paper touches upon the problem of the FR in virtual learning teams, which has been well documented and still remains a problem in collaborative learning (Hughes et al. 2015). Furthermore, it is difficult to find literature on FR in combination with technology or computer-supported tools (Kloppenburg et al. 2018).

3 Course Design

Before introducing the CLASS system, we present a brief overview of the specific educational context within which it has been developed and deployed. At our technical university, the introductory course in Human-Computer Interaction (HCI) is given twice per year, and is a compulsory course at the Computer Science and at the Media Technology study programs. It is also available as an optional course for other student groups. It is thus a large course with occasionally more than 350 students taking it simultaneously.

The course was initially primarily a theoretical course, as with many introductory HCI courses given worldwide. However, around ten years ago, a conceptual shift within the HCI discourse towards a more designerly orientation known as the 3rd wave of HCI (Bødker 2006), inspired a redesign of the course, especially regarding its assessment focus and the related assessment practices. The previous main basis for individual student assessment – a written home exam – was removed. The focus has shifted towards *collaborative practice, design critique* and *hands-on application of theory and methods*. The new course design was based on each student performing a series of group-assigned tasks as part of a larger design project, conducted in groups of 4–6 students, along with lectures, exercises, reading seminars, and project presentations in smaller divisions of 20–30 students.

The new assessment design was inspired by the peer assessment process suggested in the setup presented in Edström et al. (2005). It concerned a course with a much smaller number of students (around 30), within a radically different subject domain, i.e., Maritime Engineering, and without any specific computer system as support, it presented a promising setup for individual grading of group work including self-reflection, peer feedback, and grades - both individual and group grades - in two rounds, i.e., formative and summative assessment. In other words, we suggest that the offered set-up included the assessment elements that presented a high potential to identify free-rider in the project group work. Yet, with our large number of course participants, new computer-assisted tools to manage assessment of individual students' work in collaborative project work were needed.

The challenge was to scale up co-located *studio education*, with a specific focus on assessing not only group work but also individual student contributions in relation to the course objectives (Schön 1987). We explicitly wanted to complement and provide added value by the fact of the course being local (group work, discussions with teachers, design critique sessions). For this reason, we decided to design a completely new computer-based assessment system - based on earlier research results (e.g., Lu and Bol 2007) and the identified needs- aimed to support teachers and students, with one of the goals - to impede the FR phenomenon.

4   System Description

The CLASS system was developed and introduced in 2013 at a large university in Sweden and run for several years since that. It consists of three modules that focus on: (i) student self-reflection process, (ii) peer-grading process, and (iii) examiner feedback (Fig. 1).

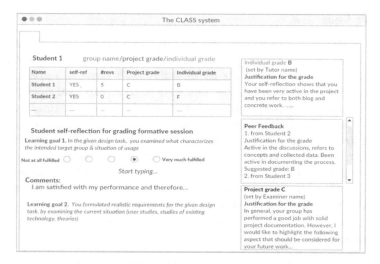

Fig. 1. The CLASS system (teacher view)

Both self-reflection and peer-grading were mandatory parts of the individual student assessment in the course, along with her/his participation in the group project. CLASS supports both teachers and students and consequently offers *a teacher view* and *a student view*. The main difference in the student view relates to the two aspects. Firstly, in contrast to the teacher view, students' peer-feedback is presented anonymously for other students in the group. Secondly, in contrast to the teacher view, a student can only see a suggested grade for the group work and his own individual grade, set by examiner, not the individual grades of his peers.

4.1 Self-reflection

To give an opportunity to a student to reflect upon her/his own work in the course, the first part of the system focuses on the individual student self-reflection in relation to the course learning objectives, as outlined in the course syllabus. Here, a student is asked to rank her/his perceived achievements towards the learning objectives of the course according to the 5-Likert scale (*Not at all fulfilled > Very well fulfilled*). Also, s/he is asked to reflectively comment (in a written form) on the chosen alternatives, according to each of the course objectives, by elaborating on her/his own contribution to the group work with some concrete examples of the individually performed work (e.g., links to the individual assignments and contributions to the group project documentation). By offering this self-assessment twice (formative and summative) during a course, students, teachers and examiners are able to continuously follow the students' individual work as well as progress, and if needed, to intervene in time. The aim is thus to help students to continuously reflect on the learning objectives of the course on the individual level. This information helps the teacher to grasp the individual student's engagement in the group work and thus contributes to her/his deeper understanding of the FR problems.

4.2 Peer Assessment

Once the deadline for submitting individual reflections has passed, the system is set into the mode of peer assessment, which as a mandatory part of student course examination. In this mode, students can read the non-anonymous self-reflections of their group members, which together with their personal experiences from the conducted project work, are used by the students to provide anonymous feedback to their peers (in a group of 5–6 students) and suggest a relevant grade for each group member. This is in line with the recent literature on FR which outlines that anonymous peer assessment of student group work is a best practice in overcoming the FR phenomenon (Kloppenburg et al. 2018). Consequently, the peer grading report included a grade given by the student along with a short justification for it. Each student had to provide separate peer grading reports for each of the group members.

4.3 Group Assessment

Using the materials presented in class and in the online documentation of each group, together with experiences from group discussions and supervision sessions, the teacher articulates a group level assessment of each group project performance and submit it to system for the examiner. The teacher goes through the online documentation of each project, accounting for students' assessment of their overall performance, and suggests a "guiding" grade for the project as a whole. The examiner then discusses the suggested grade with the teacher, examine students' work and submits the written feedback, as well as the group grade into the system. Consequently, both the course examiner and teachers have full insight into the individual students' self-reflections.

4.4 Individual Assessment

In the CLASS interface, the examiner steps through each student one by one, and sets preliminary (i.e., guiding) individual grades for the formative assessment and final grades for the summative part. The assessment of each student is presented as one summary page, consisting of teacher assessment of group work, individual self-reflection, given peer feedback and the provided by the peers feedback. This enables the examiner to skim through each individual student at a rather quick speed, and with relative ease identify who may potentially deserve a higher or lower grade than the one suggested by the overall project performance grade. Examples of situations that would elicit a lower grade than project average include: poor or incomplete personal reflection (e.g., lack of links it the group project blog that show the respective student contribution); unconstructive, impolite or incomplete feedback to peers; or lower than average performance of project work, as suggested in peer feedback. Examples that would lead to a higher grade include: ambitious and insightful self-reflection; constructive feedback to peers; and praised project performance by the peers. The way the information is presented in the system, all one summary page, these were all qualities that would quickly stand out, especially for the trained eye of the examiner familiar with the learning goals, course content and setup. Also, this way of assessment and presentation of each student contributions to the group work helps the examiner to easily identify a potential free rider in time. The examiner inserts the grade into the system, along with a short motivation. Once the assessment process is completed, students can see their preliminary grade (the formative session) or final grade (the summative session), along with examiner feedback and the comments posted by their peers.

5 Conclusions

The CLASS online system - which is planned to be offered as on open source for use by other researchers - can support three crucial aspects of CSCL: reflection, feedback and assessment, as described in Sects. 4.1–4.4. In addition, the system caters for the FR phenomenon within the CSCL context and to that end, the learning design of the CLASS system incorporates specific affordances which make it unique. In particular, the self-assessment reports of the students and the comments they receive on them subsequently from their peers helps the teacher uncover inconsistencies, especially with respect to the various levels of student contributions to the group project. Also, the different system views for the teacher and the student facilitate anonymous peer assessment in the student groups while allowing the teacher to seamlessly monitor the whole CSCL process. Our preliminary findings show that both these affordances help identify free riders in the student groups. To the best of our knowledge no research has been conducted specifically on FR in CSCL environments.

References

Bødker, S.: When second wave HCI meets third wave challenges. In: Mørch, A., Morgan, K., Bratteteig, T., Ghosh, G., Svanaes, D. (eds.) Proceedings of the 4th Nordic conference on Human-Computer Interaction: Changing Roles (NordiCHI 2006), pp. 1–8. ACM, New York (2006)

Chavez, J., Romero, M.: Group awareness, learning, and participation in computer supported collaborative learning (CSCL). Procedia - Soc. Behav. Sci. 46, 3068–3073 (2012)

Edström, K., El Gaidi, K., Hallström, S., Kuttenkeuler, J.: Integrated assessment of disciplinary, personal and interpersonal skills in a design-build course. In: Proceedings of the 1st Annual CDIO Conference, Ontario, Canada, pp. 1–9 (2005). http://www.cdio.org/files/document/file/CDIO_paper21.pdf. Accessed 15 Feb 2019

El Massah, S.S.: Addressing free riders in collaborative group work: the use of mobile application in higher education. Int. J. Educ. Manag. 32(7), 1223–1244 (2018)

Gress, C.L., Fior, M., Hadwin, A.F., Winne, P.H.: Measurement and assessment in computer-supported collaborative learning. Comput. Hum. Behav. 26(5), 806–814 (2010)

Hughes, J.E., Liu, M., Resta, P.: ICT research into K-16 teaching and learning practices. In: Educational Media and Technology Yearbook, pp. 69–82. Springer, Cham (2015)

Joyce, W.B.: On the free-rider problem in cooperative learning. J. Educ. Bus. 74(5), 271–274 (1999)

Kloppenburg, W., Nurlatifah, E., Spijkerboer, C., Yasmin, F.A.: Reducing free riding behaviour in collaborative work with computer supported tools. J. Online Informatika 3(1), 36–43 (2018)

Kobbe, L., Weinberger, A., Dillenbourg, P., Harrer, A., Hämäläinen, R., Häkkinen, P., Fischer, F.: Computer-supported collaborative learning **2**, 211–224 (2007)

Kwon, K., Liu, Y.H., Johnson, L.P.: Group regulation and social-emotional interactions observed in computer supported collaborative learning: comparison between good vs. poor collaborators. Comput. Educ. **78**, 185–200 (2014)

Lu, R., Bol, L.: A comparison of anonymous versus identifiable e-peer review on college student writing performance and the extent of critical feedback. J. Interact. Online Learn. **6**(2), 100–115 (2007)

Roberts, T.: Computer-supported collaborative learning in higher education: an introduction (2005). https://doi.org/10.4018/978-1-59140-408-8.ch001

Schon, D.A.: Educating the Reflective Practitioner. Toward a New Design for Teaching and Learning in the Professions. The Jossey-Bass Higher Education Series. Jossey-Bass Publishers, San Francisco (1987)

Slof, B., Nijdam, D., Janssen, J.: Do interpersonal skills and interpersonal perceptions predict student learning in CSCL-environments? Comput. Educ. **97**, 49–60 (2016)

Swaray, R.: An evaluation of a group project designed to reduce free-riding and promote active learning. Assess. Eval. High. Educ. **37**(3), 285–292 (2012)

Exploiting Peer Review in Microteaching Through the Ld-Feedback App in Teacher Education

Eleni Zalavra[1]([✉]), Kyparisia Papanikolaou[2], Katerina Makri[3], Konstantinos Michos[4], and Davinia Hernández-Leo[4]

[1] Athens Directorate of Secondary Education, Athens, Greece
zalavra@sch.gr
[2] School of Pedagogical and Technological Education, Athens, Greece
kpapanikolaou@aspete.gr
[3] National and Kapodistrian University of Athens, Athens, Greece
kmakrh@ppp.uoa.gr
[4] Universitat Pompeu Fabra, Barcelona, Spain
{kostas.michos,davinia.hernandez-leo}@upf.edu

Abstract. This paper presents a case study in which student-teachers practiced microteaching in the form of peer-teaching. They collaboratively developed a learning design, implemented it as a micro-lesson i.e. as a simulated teaching session and later they participated in a structured peer review activity. In particular, students initially used the Integrated Learning Design Environment to organize their task and its integrated WebCollage editor for authoring the learning design. Then these particular designs were implemented as a micro-lesson during which the authors of the design "taught" their peers who played the role of the students. After each micro-lesson, the Ld-Feedback application was used so that the student-teachers who practiced the microteaching got feedback from their peers regarding both the design and the implementation. This paper focuses on the first phase of the peer review activity explaining the use of the Ld-Feedback App and investigating the added value of involving student-teachers in creating feedback forms to evaluate their learning designs.

Keywords: Teacher education · Learning design tools · Peer review · Feedback form · Microteaching

1 Introduction

Following Schön's [1] notion of reflective practitioner, reflective practice is a seminal concept in teacher education and teacher professional development. Through reflection, teachers or student teachers question their teaching practice or design approaches, connect theory to practice and learn from their experiences. Among the reflective practices used in teacher training programs is microteaching, a simplified simulation of a regular teaching-learning process [2]. Indeed, through microteaching, researchers report that student teachers are developing their pedagogical knowledge and competence [3] as well as their reflective ability [4]. The implementation of microteaching

© Springer Nature Switzerland AG 2020
E. Popescu et al. (Eds.): MIS4TEL 2019, AISC 1008, pp. 139–147, 2020.
https://doi.org/10.1007/978-3-030-23884-1_18

entails two critical components: micro-lessons and feedback [2]. Feedback in microteaching can be provided by participants (peers, in-service teachers, tutors) who act either as observers or as students during the micro-lesson; both orally and in written form [5].

A micro lesson presupposes that student-teachers first act as designers. Learning Design (LD) calls for a fresh view on teaching as a design science, shifting the role of teachers from deliverers of pre-packaged knowledge to designers of learning [6]. In this context, the "teachers as designers" movement is lately gaining significant momentum [7] and integrates the use of LD tools for designing into teachers' practices. The involvement of teachers in the design of learning environments has been acknowledged as a form of professional development in teacher education programs. This includes various forms such as the redesign of existing activities, collaborative design in teacher teams and evidence-based adaptations [8]. Within the field of teacher education there is still room for ideas on meaningful experiences of using technology enhanced learning, specifically addressed to pre-service teachers [9].

In this paper, we present how feedback for microteaching can be organised and exploited through the Ld-Feedback App. We report on a case study with student-teachers practising microteaching aiming to provide evidence of the added value of involving this audience in creating feedback forms to evaluate their learning designs. In Sect. 2, we present the Ld-Feedback application. In Sect. 3, we describe the case study performed with student-teachers. We analyse the feedback forms they created regarding the articulation of their questions and assess their comprehension of the several types of knowledge based on the TPACK (Technology Pedagogy and Content Knowledge) framework [10]. Finally, in Sect. 4 we present the conclusions derived from the first steps of the research as well as our future plans.

2   The Ld-Feedback App

The Ld-Feedback App [11] is a web-based application designed to support feedback on learning designs. The tool is part of the Integrated Learning Design Environment (ILDE) [12], a community-oriented platform for the creation, implementation, and sharing of learning designs. The App has a friendly graphical user interface and supports feedback for learning designs in various contexts (e.g. student-to-teacher in real classroom conditions, teachers-to-teachers as co-designers of a learning design).

The functionalities for organizing, reviewing, providing and receiving feedback to learning designs are the following (see Fig. 1):

(A) Authoring of feedback forms corresponding to a learning design in ILDE. Support the teacher-designer to create a form by using two predefined types: star rating questions and open-ended questions. The teacher-designer just articulates the questions and the tool issues a link to distribute to the reviewers to answer the feedback form.

(B) Reviewing a design by completing a feedback form created by a teacher-designer. In this case, teachers or students act as reviewers of a learning design.

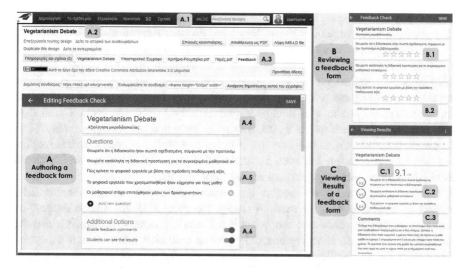

Fig. 1. Instances of the Ld-Feedback App during its utilization in the case study

(C) Viewing visualizations of feedback results corresponding to a learning design in ILDE, displayed either in the Ld-Feedback App or in ILDE.

Figure 1 shows screenshots of the Ld-Feedback App in our case study. (A) is the authoring view of a feedback form integrated into ILDE. On top of the view (A.1) is the ILDE menu in Greek (https://ilde2.upf.edu/gr/). Moving downwards, (A.2) is the title of a learning design, "Vegetarianism Debate", developed during the study. (A.3) indicates the selection of the Ld-Feedback tool for creating an associated form, among the available features. (A.4) shows the title of the feedback form, (A.5) the authoring of feedback questions and (A.6) additional options (enabling or disabling feedback comments). (B) shows the reviewers' view to a feedback form. (B.1) includes the questions authored by the teacher-designer and a star system to facilitate reviewers in rating, (B.2) allows space for the reviewers' comments. (C) shows the feedback results visualized in the Ld-Feedback App. The average value given by all reviewers in all the star-rated questions is calculated in (C.1). (C.2) shows the average given in each question, and (C.3) gives a short view of the comments provided by reviewers.

3   Case Study

3.1   Methodology

An empirical study was organized in the context of the Postgraduate program "Theory, Practice and Evaluation of Educational Practice" organized by the Faculty of Philosophy, Pedagogy and Psychology at the National and Kapodistrian University of Athens. The participants were 10 student-teachers who, during the summer semester of the academic year 2017–2018, were attending a course on Digital technologies and

Collaborative learning. Their background was from several disciplines, such as Computer Science, English or Greek language, Primary Education, etc.

The main goal of this course is to approach collaborative learning techniques not just theoretically but also practically as microteaching in the form of peer teaching. After initial theoretical groundwork, student-teachers are assigned to collaboratively develop and implement a learning design which they finally "teach" in a simulated context as a micro-lesson, with their peers playing the role of the students. To reflect on their practice, after each micro-lesson, they participate in a peer-review activity. Since the course focuses on both collaborative learning and digital technologies it is required that the learning design includes a collaborative learning technique which should ideally be implemented by applying digital technologies.

For the specific study, the tutor of the course organized the design phase in ILDE instead of a wiki previously used. Actually, the role of ILDE was to support the organization of the whole task and orient it towards a structured peer-review activity. First, student-teachers were organized in groups and each group shared a learning design in ILDE as co-editors. Then, they used the integrated tool WebCollage [13] for authoring the learning designs. The collaborative learning techniques that the designs should implement were brainstorming, debate, jigsaw and roleplay and several digital tools were selected for performing technology enhanced learning designs.

The process followed for organizing the feedback of the microteaching was applied in three phases. In the first phase, each group of student-teachers designing and performing a micro-lesson used the integrated tool Ld-Feedback App to create the feedback form of their learning design. The tutor provided relevant criteria (see Table 1) for star-rated evaluation aiming to scaffold student-teachers into formulating a feedback form based on TPACK. They were advised to follow the criteria as a baseline, but were free to articulate them as questions on their own. The feedback forms should also include an extra open-ended question allowing the reviewers to provide their comments. The second phase was implemented after each micro-lesson. Instead of printed evaluation rubrics that were previously used, each student-teacher participating as a learner in a micro-lesson used the Ld-Feedback App to fill in the feedback form assessing both the learning design and the implementation of the micro-lesson. The student-teachers who acted as teachers were able to see in the Ld-Feedback App the visualized results of their peers' review. Finally, in the third phase, each student-teacher had to think about the whole process and submit an individual reflective report addressing the peers' comments received through the Ld-Feedback App.

In this paper we analyse the feedback forms collected in the first phase to identify TPACK elements in the questions formulated by the students-teachers. Our research goal is to investigate students-teachers' comprehension of TPACK based on the elements of the micro-teaching that they address in the feedback forms.

3.2   Findings and Discussion

The main body of data was drawn from the four feedback forms created by the student-teachers' groups with the Ld-Feedback App. The forms included nine questions, each one corresponding to one of the nine evaluation criteria proposed by the tutor. We qualitatively analysed the questions of the four forms by each criterion investigating

Table 1. The criteria of the feedback forms and the TPACK knowledge addressed

Statement	Knowledge addressed
(1) Design/structure/activity flow of the micro–lesson related to the collaborative technique	Pedagogical Knowledge (PK)
(2) Appropriateness of the teaching approach	Pedagogical Content Knowledge (PCK)
(3) Added pedagogical value of the digital tool used, related to the collaborative technique	Technological Pedagogical Knowledge (TPK)
(4) Usability of the digital tool used	Technological Knowledge (TK)
(5) Usefulness of the digital tool used	Technological Pedagogical Content Knowledge (TPACK)
(6) Pedagogical skills of the teacher	Pedagogical Knowledge (PK)
(7) Classroom climate during the micro–lesson	Pedagogical Knowledge (PK)
(8) Efficient use of teaching time	Pedagogical Content Knowledge (PCK)
(9) Accomplishment of learning goals	Technological Pedagogical Content Knowledge (TPACK)

similarities and differences as well as their connection to the type of knowledge they address based on the TPACK framework [10].

The 1st question of the forms requiring feedback on the design/structure/activity flow of the micro - lesson was quite general. Three out of four groups formulated a question prompting student-teachers to assess how far or reflect if the design was adequate and aligned with the recommended bibliography of collaborative methods (e.g. "Do you find teaching properly structured based on the bibliography proposed?"). One of the groups focused on the evaluation of the structure of the collaborative method proposed, that of Jigsaw. However, one of the groups seemed to be confused asking for feedback for both the design and the implementation of the micro-lesson ("How do you evaluate the learning design? Is it consistent with the methodology suggested in the literature review? Did the implementation run smoothly?"). Thus, the particular questions are linked to the PK of student-teachers assessing their awareness of the existence of a variety of collaborative methods, each one having a specific structure.

At the 2nd question requiring feedback on the appropriateness of the teaching approach, just one of the groups formulated a very abstract question rephrasing the criterion ("Is the teaching approach appropriate/effective?"). The other three out of four groups formulated questions that required students to link the teaching approach with the learning goals of the particular topic (e.g. "Do you deem it appropriate to apply the collaborative technique to teaching the chosen topic?"). This question reflects the PCK of student-teachers focusing on the necessary matching of the teaching approach with the particular topic.

The 3rd question aims at evaluating the added value of the digital tool(s) integrated in the learning design. However, three of the groups used the term "pedagogical value" as well known, formulating the question as follows: "Did the tools used at the learning design add pedagogical value?". Only one group chose to address the particular aspects of the technology integration e.g. "Do you consider the digital tool used at the learning design to be adequate? Did it evoke the students' active engagement? Did it facilitate overcoming problems experienced in applying the collaborative technique in combination with traditional methods of teaching?". Only this last complex question links technology to pedagogy reflecting several aspects of TPK.

The 4th question focuses on the usability of the digital tool(s) integrated in the learning design. The questions formulated ranged from short ones such as "Was the digital tool used in the lesson easy for you to use?" to more personal ones such as "Was it easy for you to use the digital tools? Consider whether you could access them easily and whether you could use them on your own without requesting assistance from the teacher?". However, all of them reflect the student-teachers' TK.

The 5th question goes one step further to seek feedback about the usefulness of the digital tool(s) integrated in the design. In this question three groups tried to link the tool(s) with the support offered for the accomplishment of the activity as well as with students' engagement. The forth group took a step further to link the tools' functionality with both the topic and the didactic approach adopted. Thus, in the first approach the question reflects aspects of the TPK whilst only the last one addresses TPACK.

The next four questions refer to the implementation of the micro-lesson. The 6th question requires feedback about the teacher's skills. The questions proposed refer to several aspects of the role of the teacher such as communication skills, class management, motivating students, support offered to students and groups of students. Especially the last one focusing on student collaboration support was at the core of the questions, underlining the value of collaboration during the micro-lesson. Therefore, the particular questions are linked to the PK of students-teachers focusing on teaching skills.

The 7th question requires feedback about the class climate. The student-teachers' questions proposed concentrated on the social climate probably due to the nature of the micro-lesson. The main issue under consideration was the students' engagement as well as their interest and creativity while working on the content. Another issue considered was the role of the teacher in creating a positive climate. This question underlines specific pedagogical characteristics of a micro-lesson linked to the student-teachers' PK.

The 8th question aims to evaluate whether the student-teachers who performed the micro-lesson used the teaching time efficiently. Apart from one group using the abstract expression "Was teaching time effectively managed?", the other groups have promptly requested feedback, e.g. "Do you consider that the time distribution to the phases of the learning design was adequate? Was the time allocated to each phase sufficient during the implementation? Did the activity flow run smoothly and was the learning design implemented in time?". However, although they acknowledge time schedule as an important aspect of teaching (reflecting PK), they do not seem to link time schedule with learning outcomes investigating if and how the particular topic of the lesson was covered during the teaching time (reflecting PCK).

The 9th question investigating the accomplishment of the learning goals focused on the learning activities. The questions mostly addressed PCK as follows "Do you consider that the learning activities were appropriately designed towards accomplishing the learning objectives?" The student-teachers didn't articulate the question addressing the various elements of a learning activity that contribute to a successive outcome reflecting the complex relations among technology, pedagogy, and content. However, it seems that they acknowledge activities as the core of a student-centered lesson. Thus, it is assumed that the student-teachers as novice designers still have to enrich their skills and competences in learning design in order to comprehend the notion of "TPACK".

4 Conclusions and Future Work

In this paper, we presented a case study in the context of teacher education wherein student-teachers practiced the microteaching method in a structured multi-design phase, acting as designers, enactors, and reviewers of learning designs. Though multiple LD tools support teachers-designers at design-time, there are limited options for providing feedback on learning design implementations. We presented such a tool, the Ld-Feedback App, and how it was utilized in this study.

Findings refer to the first phase of a peer-review activity in which the student-teachers created a feedback form for the evaluation of both the design and the implementation of the micro-lesson that they performed. It is examined how they articulate the questions of the feedback form in terms of responding to the criteria given, which were based on the TPACK framework. The analysis of feedback forms focused on identifying TPACK elements in the questions formulated by the students-teachers. Moreover, these questions link the type of knowledge they expect feedback with the elements of their microteaching they expect to be evaluated. Findings show that involving student-teachers in the creation of feedback forms in accordance to the TPACK framework is beneficial. Specifically, they seem to fully acknowledge how PK is embedded in their practice in both designing and implementing collaborative techniques. Their understanding of TK regarding the integration of digital technologies also seems to be efficient. The findings are quite interesting regarding synthetic types of knowledge like PCK, TCK, TPK and TPACK. Their training in LD during their studies appears to have allowed them to comprehend PCK in terms of designing a teaching approach for a specific content or a learning goal. However, regarding the teaching approach implementation, they fail to link it with learning outcomes thus reflecting PK instead of PCK. Concerning the synthesis of technology with pedagogy (TPK), most groups managed to link the tool(s) with the support offered to the activity accomplishment as well as with the students' engagement (criteria 3 and 5). However, only one group managed to exploit the digital tools' functionality with both the topic and the didactic approach adopted reaching TPACK knowledge. In most cases, as far as TPACK is concerned, it was interpreted either as TPK or TCK.

Consequently, the findings of this study provide evidence that involving student-teachers in creating feedback forms to evaluate their micro-lessons strengthens their reflective practice. The added value of peer-review in microteaching lies in promoting the student-teachers' reflection on the various types of knowledge involved in their practice and providing them with opportunities to develop their comprehension of TPACK.

The next step of this research is to evaluate the feedback forms that the student-teachers submitted for evaluating their peers in the second phase of the case study. Analysis will investigate the student-teachers' knowledge based on TPACK framework as it is reflected on the open-ended question of their evaluation. Furthermore, we intend to investigate the student-teachers' individual reflective reports on the feedback forms they received (third phase). We plan to relate the findings of this study regarding the comprehension of the various types of knowledge with their reflection on the feedback that they received. This comparison will shed light on what types of knowledge student-teachers cultivate during such a structured review process by means of the Ld-Feedback App.

Acknowledgment. The authors acknowledge financial support for the dissemination of this work from the Special Account for Research of ASPETE through the funding program "Strengthening ASPETE's research" and the partial support of FEDER, the National Research Agency of the Spanish Ministry TIN2014-53199-C3-3-R, TIN2017-85179-C3-3-R and la Caixa Foundation 100010434. DHL is a Serra Húnter Fellow.

References

1. Schön, D.A.: Educating the Reflective Practitioner. Toward a New Design for Teaching and Learning in the Professions. The Jossey-Bass Higher Education Series (1987)
2. Yan, C., He, C.: Pair microteaching: an unrealistic pedagogy in pre-service methodology courses? J. Educ. Teach. **43**(2), 206–218 (2017)
3. Fernandez, M.L.: Investigating how and what prospective teachers learn through microteaching lesson study. Teach. Teach. Educ. **26**(2), 351–362 (2010)
4. Mergler, A.G., Tangen, D.: Using microteaching to enhance teacher efficacy in pre-service teachers. Teach. Educ. **21**(2), 199–210 (2010)
5. He, C., Yan, C.: Exploring authenticity of microteaching in pre-service teacher education programmes. Teach. Educ. **22**(3), 291–302 (2011)
6. Mor, Y., Craft, B., Hernández-Leo, D.: The art and science of learning design: editoral. Res. Learn. Technol. **21** (2013). https://journal.alt.ac.uk/index.php/rlt/article/view/1479
7. Goodyear, P., Dimitriadis, Y.: In medias res: reframing design for learning. Res. Learn. Technol. **21** (2013). https://journal.alt.ac.uk/index.php/rlt/article/view/1391
8. Voogt, J., Laferrière, T., Breuleux, A., Itow, R.C., Hickey, D.T., McKenney, S.: Collaborative design as a form of professional development. Instr. Sci. **43**(2), 259–282 (2015)
9. Papanikolaou, K., Makri, K., Roussos, P.: Learning design as a vehicle for developing TPACK in blended teacher training on technology enhanced learning. Int. J. Educ. Technol. High. Educ. **14**(1), 34 (2017)
10. Koehler, M.J., Mishra, P., Kereluik, K., Shin, T.S., Graham, C.R.: The technological pedagogical content knowledge framework. In: Handbook of Research on Educational Communications and Technology, pp. 101–111. Springer, New York (2014)
11. Michos, K., Fernández, A., Hernández-Leo, D., Calvo, R.: Ld-feedback app: connecting learning designs with students' and teachers' perceived experiences. In: European Conference on Technology Enhanced Learning, pp. 509–512. Springer, Cham, September 2017

12. Hernández Leo, D., Asensio-Pérez, J.I., Derntl, M., Pozzi, F., Chacón Pérez, J., Prieto, L.P., Persico, D.: An integrated environment for learning design. Front. ICT **5**, 9 (2018)
13. Villasclaras-Fernández, E., Hernández-Leo, D., Asensio-Pérez, J.I., Dimitriadis, Y.: WebCollage: an implementation of support for assessment design in CSCL macro-scripts. Comput. Educ. **67**, 79–97 (2013)

The Impact of Visualization Tools
on the Learning Engagement
of Accounting Students

Liana Stanca[1]([⊠]), Cristina Felea[2], Romeo Stanca[1], and Mirela Pintea[1]

[1] Business Information Systems Department, Babes-Bolyai University,
TH. Mihaly Str., 400591 Cluj-Napoca, Romania
{liana.stanca,mirela.pintea}@econ.ubbcluj.ro,
romeo_stanca@yahoo.com
[2] Faculty of Letters, Babes-Bolyai University, Horea Str.,
400591 Cluj-Napoca, Romania
cristina.felea@gmail.com

Abstract. In the big data era it is necessary for the financial-accounting education environment to redefine itself. Recently, the relevance of visualization in both pedagogical and business environments has increased, as it can support a clearer and more effective representation of information. In this context, the current paper presents a prototype course that introduces tools to visualize financial accounting concepts and was developed jointly by IT and financial teachers. The study starts with the review of specialist literature, based on which a course scenario is proposed, where processes, models and financial-accounting methods are presented with tools of modern visualization techniques developed on real-time data sets. It continues with the statistical study, which analyzes whether students' engagement increases as compared to a traditional course. The results of the statistical study confirm that visualization tools increase the degree of understanding and engagement of the financial and accounting contents by inducing students the necessary skills to identify the correct accounting decisions. We believe that the results of the paper will bring more insight into the introduction of visualization tools in the process of learning how to take accounting decisions, and we invite more academics to study, test and expand the subject of visualization in financial accounting.

Keywords: Adaptive learning · Information visualization (Dashboard) ·
Student engagement

1 Introduction

The 21st century is marked by the imposition of technology as the foundation of every field. In this period are published documents such as Europe 2020 Strategy and Partnership for 21st Century Skills according to which it is necessary to educate students in the field of digital competences. The new concepts generated by the technological revolution allow access to processing large amounts of data, hence offering the opportunity to create models, prototypes in all vital areas of education, business, etc. In

E. Popescu et al. (Eds.): MIS4TEL 2019, AISC 1008, pp. 148–156, 2020.
https://doi.org/10.1007/978-3-030-23884-1_19

this context, the concepts such as large data, data mining, visual analytics data mining, design thinking, partner education recognition 3.0, e-learning 3.0 have appeared. All these concepts affect the university and business environments, modifying the data size and thus increasing the accuracy of the results obtained by decision analysis processes. [13] states that an essential current trend for technology-based learning and teaching focuses on how to introduce analytical learning theory and practice. Starting from these developments, we have created a framework for teaching and learning the financial accounting discipline that combines entrepreneurial, smart and collaborative dimensions. In this framework, the "intelligent" dimension is introduced as learning analytics [15], which is represented in our study by theories and practices such as data analysis, agile decision making, critical visual data analysis, mining process, and design thinking. We consider it important to offer to students in economics an anchor, in-flight techniques in the new educational and professional practices generated by the era they live in, namely 21st century skills [9].

The paper is structured in four main parts starting with the literature review on e-learning, then presenting the impact of large data on the field of accounting from education to practice, accounting theory teaching scenario in big data context and statistical analysis to obtain the accounting student's profile in the big data era.

2   Literature Review

The study carried out by [13] demonstrates that the large data accountant/auditor is required to have a strong ability to identify the information and control needs of internal and external decision makers with a significant role in information governance. According to [1, 2, 11, 20], it is necessary to revise the curriculum so that accountants are given the knowledge and skills needed to conduct IT expertise and data analysis so as to be able to guide the corporate strategy by creating a link between business activities and IT functions supporting these activities. According to [10], IT applications provide the accountant with the ability to identify and report suspicious transactions, predict the quantities of products sold, customize the content of the application, and analyze user behavior in order to personalize the application and real-time analysis of social media feeds in order to approach business decision as close to what it should be. The new accounting requirements are consistent with the [1, 2] - International Accounting Accreditation Standard A7 - which directs accounting programs to develop skills and knowledge in integrating information technology into accounting and business. In [1, 2, 23], it is mentioned that there is a need for new learning experiences centered on the development of skills and knowledge related to data creation, data sharing, data analysis, data exploitation and data reporting and storage within organizations. In [14], it is stated that AIS curricula needs to be improved by introducing along with disciplines of system design, maintenance and auditing, disciplines such as server and client elements, operating systems, networking, databases, hosted software, hardware virtualization, etc.

This study aims to present a framework that combines several teaching and learning analysis processes, trying to provide a rich insight into the challenges posed by virtualization through visualization techniques of accounting phenomena based on real-time data. The transdisciplinary team who carried out the study was created with the aim of developing collaborative strategies so as to eventually set the basis for the new approach to teamwork necessary in the context of big data. Note should be made that very often there is "resistance to change" generated by fear [25] among the actors involved (students, teachers), which is followed by acceptance and use of new technologies. This was the reason why a survey of students and teachers in the fields of Economic Informatics and Accounting was conducted to investigate their attitude to the use of digital tools in the accounting field and to the development of the necessary accounting decision-making skills based on real time accounting information viewing tools. Out of the 100 participants, 36.7% said they disagreed with the idea of assimilating IT knowledge if they hold the position of accountant, while the rest of 63.3% accepted the change but claimed it would be an extra effort.

The teaching-learning framework was elaborated on [20], which states that data analysis and reporting are the objectives of the IT system and, in line with current employers' requirements; the curriculum should expose students to popular analytical and visual tools so as to increase the accountability of the accounting decision. Thus, in the authors' view, the teaching and learning process in accounting is bound to expand its boundaries so as to include theories and practices of the big data, visual analytics, and decision dashboards [12].

3 Learning Scenario for Accounting and IT in the Big Data Era

The proposed course scenario was developed on the basis of Piaget's theory [21, 22] of operational stages, taken and modified by Bruner [7, 8], which emphasized the introduction of the environmental factors and experience in the didactic act, thus contributing to the development of active and participatory methods in teaching and learning, which helps to establish a close relationship between action and thinking [12, 13] and to generate the self-regulation learning concept. Our system was designed to help students progress from training to self-training, self-evaluation and self-monitoring while developing analytical, critical, decision-making, specialty and digital skills. This approach is derived from the constructivist perspective on learning. The main features of the model are: the whole process is centered on the systemic understanding of students' individual needs and the adaptation of the teaching-learning process to these needs. During this course the authors expected to determine the best teaching tool and explanation of financial-accounting theory in the big data era, to train practitioners according to the 21st century education so as to acquire the skills needed on the labor market. The study focuses on courses hosted for the first time on faculty Moodle platform. Their structure allows students to get familiar with hypertext and multimedia comprehension strategies, which facilitates situational learning. The course design facilitates understanding by including texts and visual support (e.g. graphs, diagrams, word clouds etc.) for the following: purpose and objectives of the course, learning outcomes, review of essential notions, study questions, critical thinking on accounting problems specific for each chapter, essential notions; content is organized with fixed section headings: course notes, seminar notes, laboratory activity, project enrolment, teamwork, debates, etc. Communication and interaction are provided by asynchronous and synchronous student to student and student to teacher communication tools, discussion sheets, personal pages, project pages - all allowing for the building of a community of practice.

4 Research Methodology

The study was conducted on 473 students of the Faculty of Economics and Business Administration, during the academic years 2015–2016 and 2016–2017. In the first year of study, IT and the financial-accounting information was presented separately and students had difficulties to link IT visual tools and their economic interpretation. For our study, we considered the course on the IT processes applied in economics which were discussed without explaining the economic phenomena. In the second year, the course was delivered jointly by a computer scientist and an accountant. The IT visual tools and other technical information were presented by the computer scientist assisted by the accountant. The latter would help with explanations related to the economic phenomena. We believe that the correct understanding of economic phenomena makes it easier for students to understand the IT processes. Hence, students were offered guidance throughout the learning experience and helped with tools for visual representations of accounting phenomena such as business transactions, flowcharts and dataflow diagrams. We developed this alternative according to the A7 [1, 2] and we also relied on the results of [24].

The study hypotheses are: students in economic informatics are open to accept the design and use of visualization tools adapted to financial-accounting theory; students will be reluctant to the introduction of a section dedicated to the development and use of tools to visualize financial/accounting phenomena and hence diminish student engagement with the course. Additionally, we expect that the complex learning environment will improve digital competencies of financial-accounting students in accordance with the Digital Competence Framework for Citizens [25, 26].

Data was collected using log data to quantify student interaction with Moodle course pages through views recorded between October 2015 and February 2016 (first-year students), respectively October 2016–February 2017 (2nd year students) enrolled in the Faculty of Economics and Business Administration. Further data were collected by administering a questionnaire comprising 5 point Likert scale and 0–1 questions. Attendance sheets and final grades for student semester activity and projects were also considered.

The analysis was designed in two steps: first, dynamic page views and changes were calculated for groups with different levels of digital skills and economic knowledge within each academic year. Based on the results, the second step of the analysis compared the activity of the students with medium and advanced levels in IT and economics in the two academic years. In order to test hypotheses, we used SPSS 13.0, non-parametric Mann-Whitney tests and Kurskal-Wallis tests. The decision to apply these tests was based on the Kolmogorov-Smirnov test results. TwoStep method was applied to determine the number of clusters within the multitude of data obtained.

Descriptive analysis reveals that of the 473 students participating in the study, the majority are female (75.3%). In terms of prior knowledge of IT and economics, more than half (56.2%) have an intermediate level and the rest (43.8%) have an advanced level. As to the interest in the IT discipline, approximately a half (56.3%) show moderate interest, 16.9% high interest, and 26.8% express no interest.

In the first year of study, 35% of the subjects have low or no interest, 36.3% have a moderate interest and 28.7% have a high interest in studying IT. In what concerns computer literacy, 57.6% have intermediate level and the rest of 41.5% have a higher intermediate. The average value of the grade for the seminar activity of the participants in the study is 5.63 and the standard deviation is 2.31. The mean values of views on the months studied are: October - average view value of 0.86 and stdev of 1.263, November - average view value of 2.4 and stdev of 2.943, December - average view value of 3.364 with stdev of 2.632 and January - average view value of 3.36 and stdev of 4.127.

In the second year of study, 26.1% of students have no interest, 43% have a moderate interest and 30.9% have a high interest in the study of financial-accounting theory presented with visual tools; more than half (59.1%) have an intermediate level of computer literacy while 40.9% have a higher intermediate level. The average value of the grade for the seminar activity of participants in the study is 6.42 and the standard deviation is 2.603. The average view values for the months studied are: October - average view value of 2.34 and stdev of 1.613, November - average view value of 3.621 and stdev of 3.375, December with average view value of 4.601 with stdev of 3.976 and January with the average view value of 4.103 and stdev of 4.682.

The first step was to test for differences of views between the two academic years. According to the Mann-Whitney test result (U = 2918, p = 0.04 < 0.05) there were differences between the number of views recorded in the two years, namely the average view values recorded in the second year was higher than the view values for the first year of study. In this context, we analyzed each month to check for differences. The analysis continued with the U-Mann-Whitney test for each month, and it was found that the number of views significantly increased in the second year in October (U = 6206,000, p = 0.013), November (U = 6737.000, p = 0.03) and December (U = 6023.00; p = 0.016), which implies that the desire to engage increased from one year to the next. The comparison shows that the second-year students began to express their willingness to engage from the first month; the high interest recorded in January for both years may derive from the fact that this month is for evaluation.

The next step in the statistical analysis is the aggregation stage, during which we tested the next hypothesis: there are differences in student engagement behavior over the two years of study. The statistical results are U = 7350 and p = 0.000. On the basis of these results, it can be stated that during the second year, students' engagement rate was higher than in the first, so the interest of the students may have increased as a result of introducing the tools for visualizing financial-accounting phenomena. However, the grades obtained in the second year are lower than in the first year, which may be due to the lack of experience in dealing with processing more complex information.

We continued the statistical analysis for the second year of study by applying the TwoStep method to determine the number of clusters within the multitude of data obtained. The method indicates that in the second-year students are divided into two behavioral clusters. Clusters are characterized by the fact that the value of inter-class inertia is significantly higher than the intra-class inertia value. The set of attributes representative of the groups formed (groups of well-defined homogeneous objects) are: the number of study years of IT (F = 115,61, p = 0.000); Views (F = 14872.72, p = 0.000); Face-to-face tutorial activities attendance (F = 12912,24, p = 0.000); Participation in laboratory activities (F = 16504,06, p = 0.000); Course attendance (F = 15912,71, p = 0.000); Asynchronous/synchronous communication (F = 1064.94, p = 0.000); Agile decision skills (F = 16201,31, p = 0.000); Interpretation of tools for visual explanation of financial accounting phenomena (F = 12847.90, p = 0.000) and Final grade (F = 14624,48, p = 0,000). On the basis of these attributes, it is possible to determine the behavior of students. Thus, after applying cluster analysis, the null hypothesis was rejected, and the results demonstrate that students can be grouped into two clusters. Based on the degree of involvement measured by changes in the performance of tasks on Moodle pages (F = 281,985; p = 0.000), the analysis determined that: *the first group* includes 62% of students presenting a low to moderate interest in financial-accounting disciplines; Face-to-face tutorial activities attendance has average value (1); number of study years of IT [2–7]; attending laboratory activities between [0%–45%] times; course attendance [0%–50%] of the total number of 14; low to medium asynchronous/synchronous communication; low agile decision skills; the interpretation of tools for visual explaining of financial – accounting phenomena is mediocre and the final grade is in the interval [1; 6), which denotes that their level of knowledge at the end of the course is medium to low, and views throughout the semester are medium to high. Characteristic attributes of this cluster are: female with age between [18–22], graduate of a high school with humanities profile, did not experience an eLearning platform before university, with intermediate IT level, medium to low Moodle and course attendance activity. So, the cluster includes students with medium-to-low resistance to novelty behavior, have a reserved attitude to attending courses and work only constrained or rewarded.

The second group comprises 38% of the subjects, which present a moderate to high interest towards financial-accounting matters; face-to-face tutorial activities attendance average is higher than for the first group (1); number of study years of IT [2–9]; participation in laboratory activities between [50%–85%]; course attendance was higher than 75% of the total number of 14; medium asynchronous/synchronous communication; medium agile decision skills; the interpretation of tools for visual explaining of financial – accounting phenomena is good to very good and final grade is in the range (7–10), which denotes that their level of knowledge at the end of the course is medium to very good, and the views they have achieved throughout the semester are medium. Characteristic attributes of this cluster are: female with age between 18–34 years, high school graduates of science profile, experienced an eLearning platform before, medium to high level of IT, medium to high activity on Moodle and in face-to-face tutorial activities attendance. In conclusion, the cluster comprises students with medium-to-good behavior who have superior IT knowledge and are willing to accept challenges, present decision-making and communication abilities; they have a positive attitude to attending courses, higher critical thinking than cluster 1; their work seems to have in view labor market requirements.

The results of the cluster analysis indicate a set of attributes to determine the type of student behavior in terms of engagement to the activities of a new format of financial accounting course for the big data era. However, the results show that a relatively small number of students in economics are anchored in the realities of the labor market and accept relatively easily the introduction of tools for visualizing financial-accounting phenomena in the learning process. Overall, the findings are encouraging in the sense that acceptance of the new methods in the process of teaching the financial accounting disciplines brought about positive student reaction, represented by a higher number of students who passed the exam.

Our study has several limitations. Firstly, with regard to the research tool, the data was collected online at the end of the academic year and in a very short period of time (3 days). In the future, we intend to collect data over a longer period of time, with pre- and post- surveys. Also, the answers may have been influenced by the bonus for the semester evaluation. We believe total anonymity would have given more precise results. Self-evaluation of knowledge and skills could also influence results. Then, the target groups have distinct profiles (especially in terms of age, education, learning experience, and motivation) that have not been sufficiently analyzed against the data collected.

5 Conclusions

Our research expands the results of [5, 6] which argue that traditional configuration, in which ICT content is delivered by IT units, while reporting and analysis services are provided by accounting units, should give way to an interdisciplinary approach where the two major domains be reunited. The present research proposes a way to bring together the two fields. Our approach is validated by the results of the statistical analysis and acknowledged by the actors on the labor market; thus, 38% of the students who took part in the course during the academic year 2016–2017 agreed that under-standing how to use the IT tools and computer science concepts applied in accounting is a must, and admit that this option offers a better understanding of the accounting curriculum. Both their interest and their engagement with the course increased and, as a result, they got better acquainted with knowledge and developed skills that will further allow them to understand the role and the use of dashboards in the fields of command, control and agile economic decision. The results of our study support the following two directions: (a) the empirical results obtained by [16], namely graphical tools have a considerable positive impact on the quality of decision and do not affect the decision confidence in a performance evaluation task. Additionally, these tools also facilitate self-correction in the process of formulating financial-accounting decision; (b) the results of [3, 4] which show that supplementing tables with graphs improved the auditor's considerations in a forecasting task and of [18, 19], according to which graphs can increase the quality of the decision-making process within a balanced scorecard (BSC). In [16, 17] the importance of acquiring skills in using graphs and methods of viewing data in accounting reporting is emphasized in the context of the interdepen-dence between decision quality and confidence in decision-making. Note should be made that lack of knowledge and skills of visualization techniques, correct interpretation, and experience gained from faculty in a joint course of accounting and data science, may influence student performance. All in all, we consider that the introduction of visual tools in teaching financial-accounting decision making process will increase the future specialists' confidence in decision-making based on visualization.

References

1. Association to Advance Collegiate Schools of Business (AACSB): Information Technology Skills and Knowledge for Accounting Graduates: An Interpretation. International Accounting Accreditation Standard A7 (2014). www.aacsb.edu//media/AACSB/Publications/white-papers/accounting-accreditation-standard-7.ashx
2. American Institute of Certified Public Accountants (AICPA): Trust Services, Principles, Criteria, and Illustrations. American Institute of Certified Public Accountants, Washington, DC (2014)
3. Blocher, E., Moffie, R.P., Zmud, R.W.: Report format and task complexity: interaction in risk judgments. Acc. Organ. Soc. **11**(6), 457–470 (1986)
4. Bouquin, H.: Le controle de gestion, 5eme edn. Gestion PUF (2001)
5. Borden, J.: Education 3.0: Neuroscience Learning Research Education Technology (2016)
6. Boritz, J., Stoner, G: Technology in accounting education. In: The Routledge Companion to Accounting Education, pp. 347–375. Routledge, NY (2014)
7. Bruner, J.S.: The course of cognitive growth. Am. Psychol. **19**, 1–15 (1964)
8. Bruner, J.S.: The Culture of Education. Harvard University Press, Cambridge (1996)
9. Dede, C.: Comparing frameworks for 21st century skills. In: Bellanca, J.A., Brandt, R.A. (eds.) 21st Century Skills: Rethinking How Students Learn. Solution Tree Press, Bloomington (2010)
10. Deloitte: Tax Analytics: The Three Minute Guide (2013). http://public.deloitte.com/media/analytics/pdfs/us_ba_TaxAnalytics_091313.pdf. Accessed 30 Nov 2014
11. Deloitte: Tech Trends: Elements of Postdigital (2013). www2.deloitte.com/content/dam/Deloitte/us/Documents/technology/us-cons-tech-trends-2013.pdf
12. Dzuranin, A.C., Jones, J.R., Olvera, R.M.: Infusing data analytics into the accounting curriculum: a framework and insights from faculty. J. Acc. Educ. **43**(C), 24–39 (2018)
13. Coyne, E.M., Coyne, J.G., Walker, K.B.: Big Data information governance by accountants. Int. J. Acc. Inf. Manag. **26**(1), 153–170 (2018)
14. Coyne, J.G., Coyne, E.M., Walker, K.B.: A model to update accounting curricula for emerging technologies. J. Emerg. Technol. Acc. **13**(1), 161–169 (2016)
15. Greller, W., Drachsler, H.: Translating learning into numbers: a generic framework for learning analytics. Educ. Technol. Soc. **15**(3), 42–57 (2012)
16. Hirsch, B., Seubert, A., Sohn, M.: Visualisation of data in management accounting reports: how supplementary graphs improve every-day management judgments. J. Appl. Acc. Res. **16**(2), 221–239 (2015)
17. Horn M.: KAIST Doesn't Wait for Change in Korea, Pioneers' Education 3.0. Forbes Magazine, 17 March 2014
18. Martinsons, M., Davison, R., Tse, D.: The balanced scorecard: a foundation for the strategic management of information systems. Decis. Support Syst. **25**(1), 71–88 (1999)
19. Meyer, J.: Performance with tables and graphs: effects of training and a visual search model. Ergonomics **11**, 184–186 (2000)

20. PwC: Data driven: what students need to succeed in a rapidly changing business world. PricewaterhouseCoopers LLC, London, February 2015
21. Piaget, J.: Epistemologia genetică. Editura Dacia, Cluj Napoca (1973)
22. Piaget, J.: The Construction of Reality in the Child. Basic, New York (1954). https://doi.org/10.1037/11168-000. Trans. M. Cook
23. Romney, M.B., Steinbart, P.J.: Accounting Information Systems. Pearson Education, London (2015)
24. Yigitbasioglu, O., Velcu, O.: A review of dashboards in performance management: implications for design and research. Int. J. Acc. Inf. Syst. **13**, 41–59 (2012)
25. Conole, G., Alevizou, P.: A literature review of the use of Web 2.0 tools in higher education. A Report Commissioned by the Higher Education Academy. The Open University Walton Hall, Milton Keynes (2010). https://core.ac.uk/download/pdf/5162.pdf
26. Ferrari, A.: Digital Competence in Practice: An Analysis of Frameworks. Publications Office of the European Union (2012). ftp.jrc.es/EURdoc/JRC68116.pdf

Correction to: Methodologies and Intelligent Systems for Technology Enhanced Learning, 9th International Conference, Workshops

Elvira Popescu, Ana Belén Gil, Loreto Lancia,
Luigia Simona Sica, and Anna Mavroudi

Correction to:
E. Popescu et al. (Eds.): *Methodologies and Intelligent Systems*
for Technology Enhanced Learning, 9th International
Conference, Workshops, **AISC 1008,**
https://doi.org/10.1007/978-3-030-23884-1

In the original version of the book, the following belated corrections in the chapters have been incorporated:

1. In chapter 8, the author names Berardi Anna, Galeoto Giovanni, Tofani Marco, Mangone Massimiliano, Ratti Serena, Danti Arianna, Sansoni Julita, Marquez Maria Auxiliadora have been changed to Anna Berardi, Giovanni Galeoto, Marco Tofani, Massimiliano Mangone, Serena Ratti, Arianna Danti, Julita Sansoni and Maria Auxiliadora Marquez.
2. In chapter 9, the author names Salviani Silvia, Tofani Marco, Fabbrini Giovanni, Leo Antonio, Berardi Anna, Sansoni Julita, Galeoto Giovanni have been changed to Silvia Salviani, Marco Tofani, Giovanni Fabbrini, Antonio Leo, Anna Berardi, Julita Sansoni and Giovanni Galeoto.

The corrected chapter has been updated with the changes.

The updated version of these chapters can be found at
https://doi.org/10.1007/978-3-030-23884-1_8
https://doi.org/10.1007/978-3-030-23884-1_9

Author Index

© Springer Nature Switzerland AG 2020
E. Popescu et al. (Eds.): MIS4TEL 2019, AISC 1008, pp. 157–158, 2020.
https://doi.org/10.1007/978-3-030-23884-1

Printed in the United States
By Bookmasters